ENERGY SCIENCE, ENGINEERING AND TECHNOLOGY

# WASTE HEAT

## SOURCES AND USES

# ENERGY SCIENCE, ENGINEERING AND TECHNOLOGY

Additional books and e-books in this series can be found on Nova's website under the Series tab.

ENERGY SCIENCE, ENGINEERING AND TECHNOLOGY

# WASTE HEAT

## SOURCES AND USES

MICHÈLE PICARD
EDITOR

nova science publishers
New York

**Library of Congress Cataloging-in-Publication Data**

ISBN: 978-1-53617-355-0

*Published by Nova Science Publishers, Inc. † New York*

# CONTENTS

# PREFACE

Waste Heat: Sources and Uses introduces the main waste heat sources that come from the industrial, electric, commercial, residential and transportation sectors of society. The wide applications of waste heat through heat pumps are then detailed, mainly including gas heating, hot water production, and steam production (for sterilization, distillation, rectification, etc.).

A summary of the main findings and trends in the area of ejector-enhanced efficiency of waste heat recovery cycles and systems using low boiling point working fluids. Following a brief overview of the ejector, selected configurations of interest are presented and discussed in a variety of applications.

The purpose of the following chapter is to provide a concise overview of waste heat availability and energy capturing potential, and the different technologies available for its recovery as well as their advantages and challenges.

In closing, the different types of waste heat recovery technologies that are used in cement plants to produce electricity are reviewed, and the potential for electricity production from heat exhaust gases of three cement plants by using waste heat recovery technology is discussed.

Chapter 1 - Waste heat can come from any activity of humans, i.e., cement production, steel making, power generation, ceramics sintering, etc., and it can account for up to 60% of the total thermal energy produced. Heat pump is a system that can efficiently transfer heat (or thermal energy) from a low-temperature body to a high-temperature body, and it can be efficiently used to recover and/or upgrade waste heat for wide applications. This chapter initially introduces the main waste heat sources that come from the industrial, electricity, commercial, residential and transportation sectors of society. The wide applications of waste heat through heat pump are then detailed, and these mainly include gas heating (for drying, preheating, electricity generation, etc.), hot water production (for hot water supply, space heating, snow melting, etc.), and steam production (for sterilization, distillation, rectification, etc.). Main conclusions and perspectives are also briefly stated and summarized. With the rich sources of waste heat and the advantages of heat pump, heat pump has a bright future in recovery of waste heat that can effectively narrow the energy gap and reduce negative environmental impacts of our energy production and application practice.

Chapter 2 - Energy consumption for industrial, commercial and domestic needs continues to increase and supplies to satisfy this demand with low environmental impact represents a major challenge. At the same time, considerable amounts of energy input to industrial activities are released in the environment at low temperature. Waste Heat Recovery (WHR) is identified among the most appropriate solutions to this issue, however available waste heat literature shows that technologies to extract and use low-grade waste heat effectively are scarce.

In the recent literature devoted to WHR activities, the ejector technology has been receiving particular attention, due to its suitability to harness waste heat at the lower range of temperatures where other established technologies are no longer efficient.

The use of ejectors as low temperature waste heat transformers to produce cooling and compression effects has been abundantly discussed in the refrigeration and heat pumping literature. Integrated in WHR systems in order to build various forms of combination with other equipment, this device has gained renewed interest as a solution for low temperature heat recovery to enhance overall efficiency, reduce costs and control greenhouse gases (GHG) emissions. Uses include cooling and heating operations, industrial process enhancement as a heat exchanger or integrated in systems to improve their overall efficiency as a circulating, expander, pumping or compression agent.

In the present chapter, the ejector potential to enhance a variety of conventional technologies for waste heat recovery has thus been the main objective, especially at decreasing waste heat temperature grades. The focus is on the ejector use as a component in processes, in cycles to form hybrid units or integrated in systems as a device to improve their operation and performance. The result is a summary of the main findings and trends, in the area of ejector-enhanced efficiency of WHR cycles and systems using low boiling point working fluids. Following a brief overview presentation of the ejector, selected configurations of interest, are presented and discussed in a variety of application domains according to three main topics: refrigeration-heating systems, power systems and tri-generation systems.

Chapter 3 - In most industrialised countries, energy consumption is high and the demand, mainly based on fossil resources for its production, is increasing. A recent editorial document reports that industrial needs for heat amount approximately to 70% of the total demand and generate 33% of overall greenhouse gases (GHGs). In addition, up to 50% of energy input to processes end up as waste heat in hot exhaust gases, cooling water, or from equipment surfaces and heated products.

It is nowadays very clear that recovering a portio of this heat for recycling in the processes themselves or for alternative forms and uses will substantially reinforce economic and environmental benefits. For example, recovered waste heat may contribute to electricity generation, preheat combustion air or boiler water as well as produce cooling and heating effects, depending on particular needs.

Efforts to use efficiently the available energy with minimal losses have thus been ongoing for a long time; research and development on related issues has been prolific. Methodologies and approaches for optimized waste heat recovery, where source-sink compatibility, temporal availability and adapted technology selection continue to be proposed. Applied research activities concern all aspects open to improvement such as waste heat characterization (type, quality, quantity, etc...), use of conventional technologies, and development of new concepts, depending on the need, the kind of process, including economic considerations.

Waste heat recovery at high temperature, at different levels, may be transferred with passive heat exchangers between streams and processes. In addition it can produce power by means of conventional steam turbines or ORCs. With decreasing temperatures, waste heat recovery becomes progressively limited in terms of energy harnessing and efficiencies. In this range of temperatures, alternative technologies and systems for temperature uplift, refrigeration and space heating are more frequently proposed for investigation.

The purpose of this chapter is to provide a concise overview of waste heat availability and energy capturing potential, as well as the different technologies available for its recovery with their advantages and challenges. In view of their importance, selected conventional and emerging technologies such as Rankine cycle, ORCs and heat pump types will be discussed in some detail. Steam ejectors are introduced in the context of specific applications where they have been used for many years and are available in various sizes and geometries. Ejector technology extended to a range of other working fluids make it suitable to harness waste heat at the lower ranges of temperature, which will receive particular attention in a separate chapter.

Chapter 4 - The cement sector is one of the most-energy using industries around the world. Generally, 75% of the used energy in the cement sector is thermal energy, and 25% is electrical energy. The considerable part of thermal energy, which is consumed to produce clinker in cement plants, is wasted from stacks in the form of convection. A waste heat recovery (WHR) system would be utilized to raise the efficiency of a cement plant; hence, it could reduce greenhouse gas emissions. Besides, it could decrease the exhaust gas temperature and the amount of waste heat to the atmosphere. In this chapter, first, the different types of WHR technologies that are used in cement plants to produce electricity were reviewed. Then, the potential for electricity production from heat exhaust gases of three cement plants by using WHR technology was discussed. Finally, the economic feasibility of building a WHR unit in these plants was evaluated.

In: Waste Heat: Sources and Uses
Editor: Michèle Picard
ISBN: 978-1-53617-355-0


*Chapter 1*

# USES OF WASTE HEAT VIA HEAT PUMP

***Yaning Zhang[1,*], Wenke Zhao[1], Lei Li[1], Wentao Su[2], Bingxi Li[1,†] and Xin Chen[1]***

[1]School of Energy Science and Engineering, Harbin Institute of Technology, Harbin, China

[2]College of Petroleum Engineering, Liaoning Shihua University, Fushun, China

## ABSTRACT

Waste heat can come from any activity of humans, i.e., cement production, steel making, power generation, ceramics sintering, etc., and it can account for up to 60% of the total thermal energy produced. Heat pump is a system that can efficiently transfer heat (or thermal energy) from a low-temperature body to a high-temperature body, and it can be efficiently used to recover and/or upgrade waste heat for wide applications. This chapter initially introduces the main waste heat sources that come from the industrial, electricity, commercial, residential and transportation sectors of society. The wide applications of waste heat through heat pump are then detailed, and these mainly include gas heating (for drying, preheating,

* Corresponding Author's Email: ynzhang@hit.edu.cn.
† Corresponding Author's Email: libx@hit.edu.cn.

electricity generation, etc.), hot water production (for hot water supply, space heating, snow melting, etc.), and steam production (for sterilization, distillation, rectification, etc.). Main conclusions and perspectives are also briefly stated and summarized. With the rich sources of waste heat and the advantages of heat pump, heat pump has a bright future in recovery of waste heat that can effectively narrow the energy gap and reduce negative environmental impacts of our energy production and application practice.

**Keywords:** waste heat, heat pump, sources, applications, gas heating, hot water production, steam production

# 1. Introduction

Waste heat can be narrowly defined as the energy dissipated from the energy input to the ambient in different forms during industry processes [1], it can also be broadly defined as the thermal energy that all energetic losses eventually get converted to [2]. The broad definition of waste heat is adopted in this study, actually, waste heat derives from many human activities, such as cement production, steel making, power generation, ceramics sintering, residential supply, commercial service, and traffic transportation [3], and it accounts for more than 60% global primary energy consumption [4]. Therefore, the usage of waste heat is crucial to save primary energy and reduce $CO_2$ emission [5], and it can be divided into direct usage and heat conversion [6]. Direct uses are simple, and they mainly extract the waste heat by economizer [7], passive air preheater [8], plate heat exchanger [9], and so on, whereas they have several disadvantages: (a) the temperature of direct usage is lower than the temperature of waste heat, and (b) the direct usage is mainly focused on industry processes [10]. Heat conversion can convert the waste heat to different forms including electricity, steam, and hot water for the users, and the cycles or systems can be organic Rankine cycle (ORC) [11], Kalina cycle [12], and heat pump [13], etc.

One promising technology for waste heat exploitation is the heat pump cycle, because heat pump has extensive applications in heat conversion, such as gas heating, hot water and steam production [13] with the advantages of low/no $CO_2$ emission, high energy efficiency, and various heat sources/sinks

[14]. Heat pump adopts the low-grade medium (i.e., air, water, ground) and then produces high-grade medium (i.e., heating gas, hot water, saturated steam) [15]. The highest temperature at present that a heat pump can obtain is about 160°C [16], and the temperature is still increased to receive more extensive applications [17]. Most modern heat pumps are vapor compression systems or vapor absorption systems [18], applied in all economic sectors including industry, transport, and construction [19]. Other types of heat pumps include sorption heat pump [20], chemical heat pump [21], membrane heat pump [22], and so on, and they have been developed to satisfy special applications or improve operation efficiencies.

A typical vapor compression heat pump mainly consists of a compressor, a condenser, an evaporator, and an expansion valve [23], and the heat pump can be divided into air source heat pump [24], ground source heat pump [25], and water source heat pump [26] according to the heat sources, and the compressor is the key component to drive the refrigerant in the cycle by consuming electrical energy. Various optimized compression cycles have been proposed to improve system efficiencies, such as internal heat exchanger cycle [27], ejector-expansion cycle [28], ejector-expansion with internal heat exchanger cycle, expander cycle, expander with internal heat exchanger cycle and oil-flooded compression cycle [29].

Vapor absorption heat pumps are not dependent on electricity whereas they are limited to large industrial applications or district heating/cooling plants as compared with vapor compression heat pumps [30]. There are two types of vapor absorption heat pump, one is upgrading the medium at low-temperature level to the middle-temperature level through consuming thermal energy at high-temperature level (common forms: single effect [31] and double effect [32]), and the other is utilizing the heat at middle-temperature level to produce energy at high-temperature level and the discharge is at low-temperature level (this type is also called heat transformer). The common components of the two types of vapor absorption heat pumps are generator, absorber, evaporator, condenser, and solution heat exchanger [33].

The vapor compression heat pump and vapor absorption heat pump also have been combined with other technologies to improve system performances and reduce operational cost. These hybrid systems include solar-assisted vapor compression heat pump for improving performance [34], vapor compression heat pump with organic Rankine cycle for producing electricity and heat [35], reversible absorption heat pump with an internal combustion engine for producing heating and cooling [36], hybrid absorption-compression heat pump for producing high-temperature (>130 °C) saturated steam [37], and so on.

Recovering waste heat by heat pumps can improve energy utilization efficiency and reduce air pollution emissions. In this chapter, the sources of waste heat are presented, then the wide applications of waste heat through heat pump are detailed. The wide applications mainly include gas heating (for drying, preheating, electricity generation, etc.), hot water production (for hot water supply, space heating, snow melting, etc.), and steam production (for sterilization, distillation, rectification, etc.).

## 2. Sources of Waste Heat

Waste heat can be classified into three levels according to the temperatures: (a) high-temperature waste heat (≥300°C), (b) middle-temperature waste heat (100 ~ 299°C), and (c) low-temperature waste heat (<100°C), with the proportions of 21%, 16%, and 63% in the world, respectively [3]. There are five main sectors in energy consumption: industrial, electricity, commercial, residential, and transportation sectors [38], and the sectoral proportions of waste heat distribution are shown in Table 1.

**Table 1. Sectoral proportions of waste heat distribution [3]**

| Sector | Global waste heat (PJ) | Proportion of different waste heat temperature (%) | | |
|---|---|---|---|---|
| | | High | Middle | Low |
| Industrial | 31.90 | 38 | 20 | 42 |
| Electricity | 107.21 | 0 | 12 | 88 |
| Commercial | 14.26 | 36 | 5 | 59 |
| Residential | 30.42 | 0 | 64 | 36 |
| Transportation | 61.93 | 54 | 0 | 46 |

## 2.1. High-Temperature Waste Heat

The total quantity of high-temperature waste heat is about 50.696 PJ, which is distributed in the transportation, industrial, and commercial sectors, with the proportions of 63%, 24%, and 10%, respectively. The typical recovery technologies for high-temperature waste heat are the regenerative burner, recuperative burner, regenerator, recuperator, steam Rankine cycle, Kalina cycle, waste heat boiler, and thermoelectric generation [6, 39-41].

The regenerative and recuperative burners were applied to capture and utilize the hot flue gas to preheat the inlet air, and these burners can reduce fuel consumption and improve thermal efficiency. For example, the preheating air temperatures for the regenerative and recuperative burners can reach 1000 °C and 700 °C when the furnace capacities were 40 ~ 150 tons/h, respectively [42]. The regenerator and recuperator are ceramic or metallic heat exchangers, which were used in the filtration combustion VOC (volatile organic components) oxidizing reactor to recover the high-temperature exhausts (800 ~ 1100°C) and reduce the polluted emissions [43]. The steam Rankine cycle uses water/steam as the working fluid, which was applied to recover the exhausts (temperature of 509 °C) in a heavy-duty diesel engine to increase the plant efficiency [44]. The Kalina cycle is based on the ammonia-water mixture, Larsen et al. [45] used it for the exhaust heat recovery from large marine engines with a heat source temperature of 346 °C. The waste heat boiler recovers the exhausts to generate steam, Liu et al. [46] applied it to diesel engines of offshore oil production facilities with waste heat temperature in the range of 350 ~ 400 °C. The thermoelectric generator can recover the waste heat from automotive exhaust and diesel generator exhaust with heat source temperatures of 400 ~ 700 °C [47] and 500 °C [48]. The high-temperature waste heat sources and recovery technologies are listed in Table 2.

**Table 2. High-temperature waste heat sources and recovery technologies [44-50]**

| No. | Waste heat source | Temperature range (°C) | Recovery technology |
|---|---|---|---|
| 1 | Exhausts from electric arc furnace | 1500 ~ 1700 | Preheater, bucket types scrap preheater, twin-shell furnace, Fuchs shaft furnace, and evaporative cooling technology |
| 2 | Basic oxygen process in the glass industry | 1250 ~ 1700 | Boiler/HRSG (Heat Recovery Steam Generator), basic oxygen process gas cooling device, and dedusting device |
| 3 | Regenerative glass system | 400 ~ 600 | Stationary regenerator, cullet preheater, HRSG, and organic Rankine cycle |
| 4 | Oxy-fuel glass system | 1450 ~1550 | Batch/Cullet preheater, HRSG, and organic Rankine cycle |
| 5 | Melting furnaces (fuel fired) in the Aluminum industry | 750 ~ 950 | Conventional recuperator, regenerative burner, charge preheater, and HRSG |
| 6 | Anode baking in the Aluminum industry | 300 ~ 500 | Thermoelectric generator, sidewall heat exchanger, and air preheater |
| 7 | Calcining for cement | 300 ~ 500 | Cyclone and fluidized bed alumina cooler |
| 8 | High-temperature solid waste | 650 ~ 1000 | Air preheater |
| 9 | Incinerator | 650 ~ 1450 | Steam Rankine cycle |
| 10 | Exhausts from combustion gas turbine | 370 ~ 540 | Organic Rankine cycle |
| 11 | Automotive exhausts | 400 ~700 | Thermoelectric generator |
| 12 | Engine exhausts | 346 ~ 509 | Steam Rankine cycle, Kalina cycle, waste heat boiler, and thermoelectric generator |

## 2.2. Middle-Temperature Waste Heat

The total quantity of middle-temperature waste heat is about 39.429 PJ, and it is distributed in the residential, electricity, industrial, and commercial sectors, with the proportions of 49%, 33%, 16%, and 2%, respectively. The typical recovery technologies for middle-temperature waste heat are the organic Rankine cycle, Kalina cycle, phase change material, and heat pipe exchanger [6, 39-41].

Thekdi et al. [51] suggested that the suitable recovery temperature of waste heat source was 95 ~ 260 °C by the organic Rankine cycle. Mamun et al. [52] explored the feasibility of waste heat (285 °C) recovery by the organic Rankine cycle in the steel mills. Milewski et al. [53] reported that the Kalina cycle could efficiently recover the waste heat in the steel industry when the temperature of waste heat was above 200 °C. Kostowski et al. [54] applied the phase change material to recover the exhaust (above 120 °C) in a natural gas compressor station. Tian et al. [55] proposed a new heat pipe exchanger to recover the waste heat (around 175 °C) in the dyeing and printing industry. The middle-temperature waste heat sources and recovery technologies are listed in Table 3.

**Table 3. Middle-temperature waste heat sources and recovery technologies [6, 49, 52-55]**

| No. | Waste heat source | Temperature range (°C) | Recovery technology |
|---|---|---|---|
| 1 | Blast furnace gases | 200 ~ 300 | Scrubber, top gas pressure recovery turbine, and recuperator hot blast stove |
| 2 | Waste heat in the steel industry | 200 ~ 300 | Organic Rankine cycle and Kalina cycle |
| 3 | Exhausts in the natural gas compressor station. | 120 ~ 300 | Phase change material |
| 4 | Waste heat in the dyeing and printing | ~ 175 | Heat pipe exchanger |
| 5 | Drying, baking, and curing ovens | 100 ~ 230 | Organic Rankine cycle and space heating |
| 6 | Cooling water from annealing furnaces | 100 ~ 230 | Domestic water heating, organic Rankine cycle and space heating |
| 7 | Exhaust gases exiting recovery devices in gas-fired boilers, ethylene furnaces, etc. | 100 ~ 230 | Domestic water heating, organic Rankine cycle and space heating |

## 2.3. Low-Temperature Waste Heat

The total quantity of low-temperature waste heat is about 155.592 PJ, and it is distributed in the electricity, transportation, industrial, residential, and commercial sectors, with the proportions of 61%, 18%, 9%, 7%, and 5%, respectively. The typical recovery technologies for low-temperature waste heat are the plate heat exchanger, shell and tube heat exchanger, and heat pump [6, 39-41].

**Table 4. Low-temperature waste heat sources and recovery technologies [50, 56-67]**

| No. | Waste heat source | Temperature range (oC) | Recovery technology |
|---|---|---|---|
| 1 | Process steam condensate cooling water | 50 ~ 90 | District heating |
| 2 | Cooling water from the internal combustion engine | 66 ~ 100 | Heating domestic water |
| 3 | Bearing heating | 32 ~ 88 | Kalina cycle |
| 4 | Air compression | 27 ~ 50 | Piezoelectric device |
| 5 | Welding and plastic injection machines | 32 ~ 88 | Adsorption and absorption refrigeration cycle |
| 6 | Waste heat in food, tobacco, and paper industries | 20 ~ 70 | Asymmetric plate heat exchanger |
| 7 | Sewage | 8 ~ 32 | Heat pump |
| 8 | Wastewater | 40 ~ 70 | Heat pump |
| 9 | Flue gas and cooling water from thermoelectric power plants | 40 ~ 100 | Membrane distillation |
| 10 | Kiln gasses and cooling streams in the cement and ceramic industries | 50 ~ 80 | Membrane distillation |
| 11 | Greywater streams in the buildings | 36 ~ 65 | Heat pump and heat exchanger |

Palomba et al. [56] stated that the low-temperature waste heat in food, tobacco, and paper industries at the temperature range of 20 ~ 70 °C can be utilized by a new asymmetric plate heat exchanger. Shen et al. [57] explored the performances (heat recovery and water purification) of the sewage-resource-based system with sewage temperature in the range of 8 ~ 32 °C. Kahraman et al. [58] studied the performance of a wastewater source heat pump for heat source temperature of 40 °C. Wang et al. [59] reported the oily wastewater recovery in the union station at the temperature of 40 ~ 70 °C by the absorption heat pump technology. Gustafson et al. [60] reported that the membrane distillation can apply for intermittent (i.e., flue gas and cooling water from thermoelectric power plants at 40 ~ 100 °C [61]) and variable-temperature (kiln gasses and cooling streams in the cement and ceramic industries at 50 ~ 80 °C [62, 63]) waste heat to produce water and store heat. Bertrand et al. [64] investigated the utilization of various greywater streams (36 ~ 42.5 °C in the bathtub [64], 45 ~ 65 °C in the kitchen [65] and 37 ~ 49 °C in the laundry [66, 67]) by wastewater heat recovery systems inside buildings. The low-temperature waste heat sources and recovery technologies are summarized in Table 4.

## 2.4. Conclusion

The high and middle-temperature waste heats are easier and more feasible to be recovered as compared with the low-temperature waste heat because they have high energy density and they are multitudinous transformation forms [10]. However, some certain high and middle-temperature waste heats are not recovered due to the corrosive or fouling chemicals and economic efficiency (e.g., small metal casting and glass operations) [6].

The amount of low-temperature waste heat is enormous, its exploitation and utilization are significantly important to reduce pollution emissions and improve energy efficiency. The main recovery option, heat pump, will be introduced in the following sections.

# 3. Waste Heat Used by Heat Pump for Gas Heating

## 3.1. Drying

The drying process has been widely applied in agricultural, biological and wood products [68]. Heat pump dryer is an attractive process due to its economic efficiency [69] and good quality [70]. The drying temperature is between 30 ~ 85 °C, which enhances the heat transfer rate in the drying materials and moisture diffusion rate in the ambient air [71].

Figure 1 shows the drying schematic system driven by a vapor compression heat pump. Fresh air is heated by the condenser and then is transported to the drying chamber. The exhaust gas from the chamber is mixed with waste heat and then it is circulated to the evaporator. Taşeri et al. [72] investigated the drying kinetics and quality parameters of grape pomace at the drying air temperature of 45 °C by a vapor compression heat pump. The condenser was the heat source to heat the fresh air, and the evaporator absorbs the energy from the mixture of 75% exhaust air and 25% fresh air.

Liu et al. [73] analyzed the drying performance of fresh carrot chips at the drying air temperature of 30 ~ 40 °C by a modified vapor compression heat pump. The air was dehumidified at the airside of the evaporator, and an auxiliary condenser, an air control valve (ACV) and a bypass duct were added in the drying system to improve the control accuracy and drying quality. The system is shown in Figure 2.

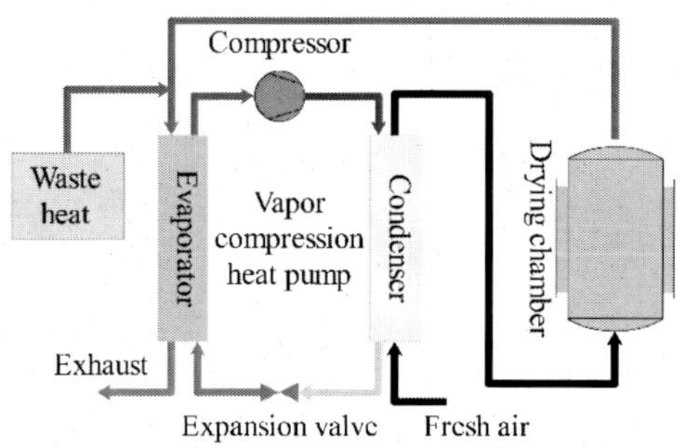


Figure 1. Open cycle drying system based on the vapor compression heat pump. Modified from [72].

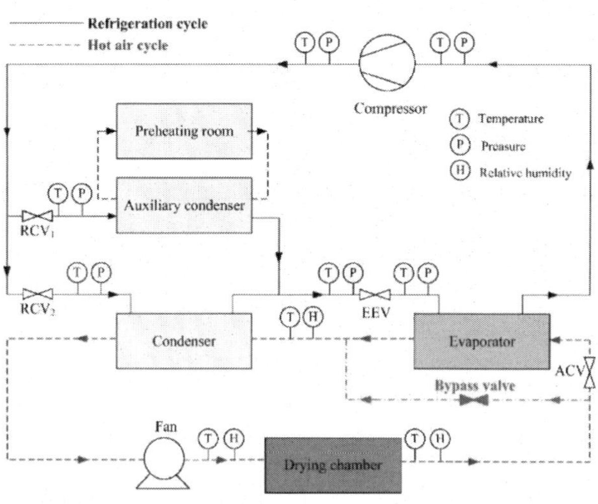


Figure 2. Close cycle drying system based on the modified vapor compression heat pump [73].

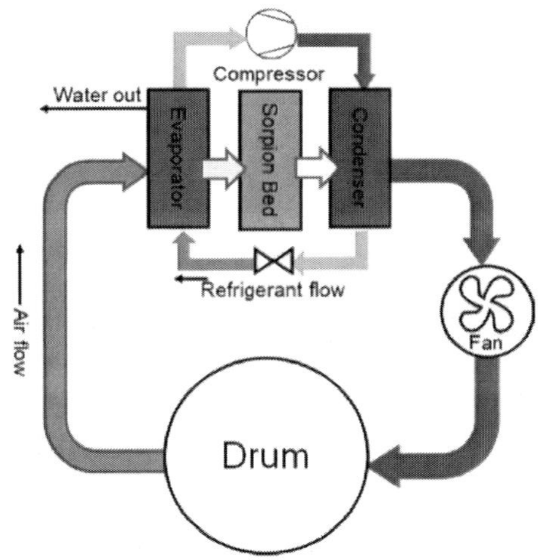


Figure 3. Close cycle drying system based on vapor compression heat pump and sorption bed [74].

Cranston et al. [74] proposed a hybrid sorption/heat pump washer-dryer system to dry the textiles at the drying air temperature of 45 °C, and the system is shown in Figure 3. The textiles were placed in the drum, which is an ionic-liquid sorption bed to dehumidify the moisture in the circulating air.

## 3.2. Preheating

Preheating for the raw materials, combustion air, and so on, is common in the industries [6]. This process utilizes the waste heat to increase the temperature of the raw materials or combustion air thereby improving the efficiency of the whole system. When the temperature of waste heat is not enough to preheat a material, the heat pump rises the temperature to ensure the preheating process.

Figure 4 shows the preheating system based on the vapor compression heat pump. The main stages include combustion, drying, heating, and so on. Yahya et al. [75] utilized the solar collector and biomass furnace to heat the gas preheated by the vapor compression heat pump, and the heated gas reached 79 ~ 85 °C when the gas entered the fluidized bed.

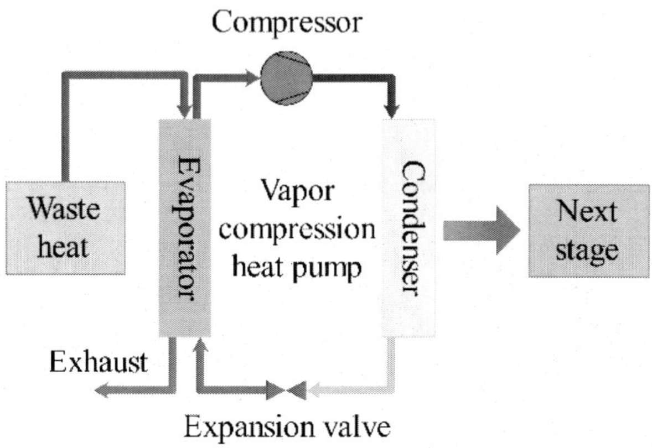


Figure 4. Preheating system based on the vapor compression heat pump. Modified from [75].

### 3.3. Electricity Generation

Electricity generation system using heat pump are generally operated in two ways: (a) vapor compression heat pump is combined with organic Rankine cycle [76] (as shown in Figure 5), and (b) vapor absorption heat pump is combined with microturbine [77] (as shown in Figure 6).

The organic Rankine cycle is a promising technology to convert low and middle-temperature waste heat into electricity [78], and it is integrated with a vapor compression heat pump that can improve the overall net power output [76]. The system was integrated by two parts (as shown in Figure 5): (a) the condenser in the heat pump was worked as an evaporator in the organic Rankine cycle to upgrade the fluid temperature of organic Rankine cycle, and (b) the waste heat was recovered by another evaporator in the organic Rankine cycle, then was transported to the evaporator in the heat pump, and eventually returned to the another evaporator in the organic Rankine cycle, this process took full advantage of utilizing the waste heat.

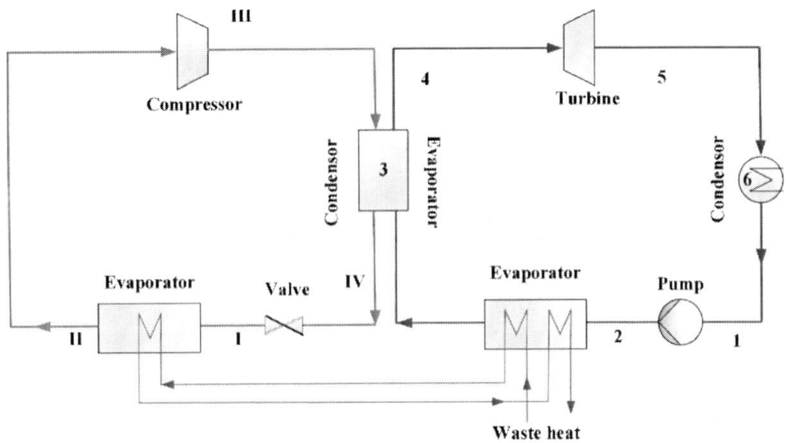


Figure 5. Electricity generation system integrated by vapor compression heat pump and organic Rankine cycle [76].

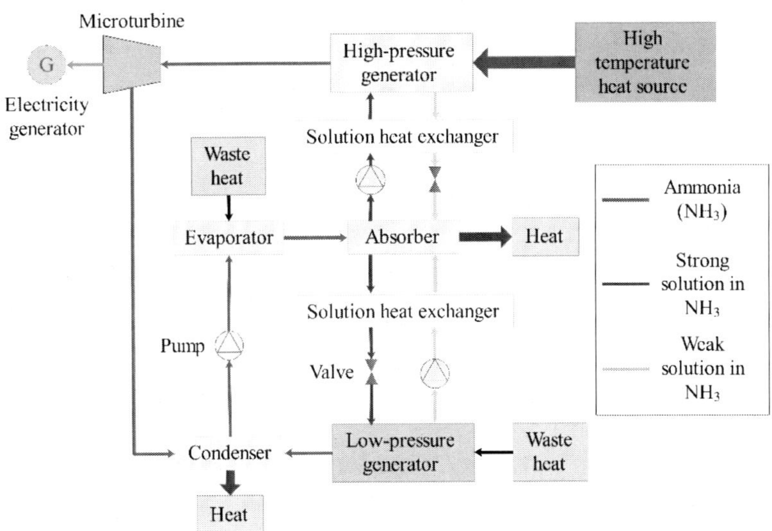


Figure 6. Electricity generation system integrated by vapor absorption heat pump and microturbine. Modified from [77].

The electricity generation system is an absorption heat pump combined with microturbine, as shown in Figure 6. The high-temperature heat source was supplied to the high-pressure generator to produce ammonia ($NH_3$) at high temperature and pressure for the microturbine, and the temperature of the heat source was more than 100 °C. The waste heat for the evaporator and low-pressure generator was the same, and the temperature of waste heat was lower than 100 °C. This electricity generation system can also generate heat, one was the main heat from the absorber, and the other was a small amount of heat from the condenser. This system can be used in the pulp and paper mill to recover waste heat [77, 79], the high-temperature heat source and waste heat were surplus steam of 155°C and black liquid of 70 °C, respectively, and the productions were the electricity of 0.21 MW and water of 110 °C, respectively.

### 3.4. Conclusion

Gas heating by heat pump can be used for drying, preheating, and electricity generation in the fields of food, paper, textiles, non-metallic minerals, non-ferrous metal, steel, and wood products, with the gas temperatures of 30 ~ 150 °C [80]. The types of heat pump applied for gas heating are vapor compression heat pump, vapor absorption heat pump, and vapor compression heat pump with organic Rankine cycle, where the most widely used type is vapor compression heat pump due to its advantages of reliability, maturity, and safety.

## 4. Waste Heat Used by Heat Pump for Hot Water Production

### 4.1. Hot Water Supply

Hot water is essential for daily life and industry production, and heat pump is a convenient technology to supply hot water and reduce fuel consumption [81]. There are two main methods to supply hot water: (a) vapor compression heat pump [82, 84], and (b) vapor absorption heat pump [85].

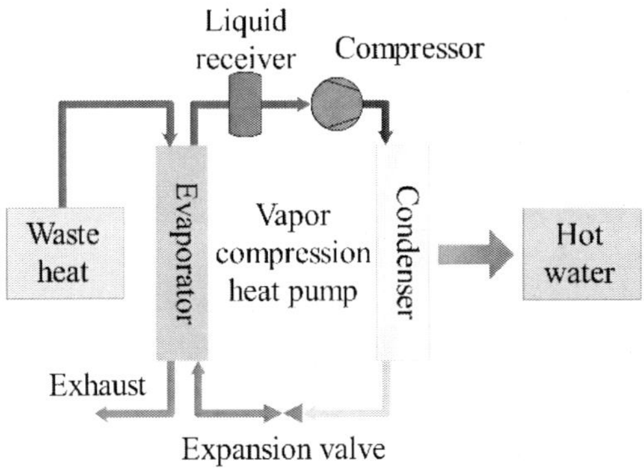


Figure 7. Hot water system by vapor compression heat pump. Modified from [82].

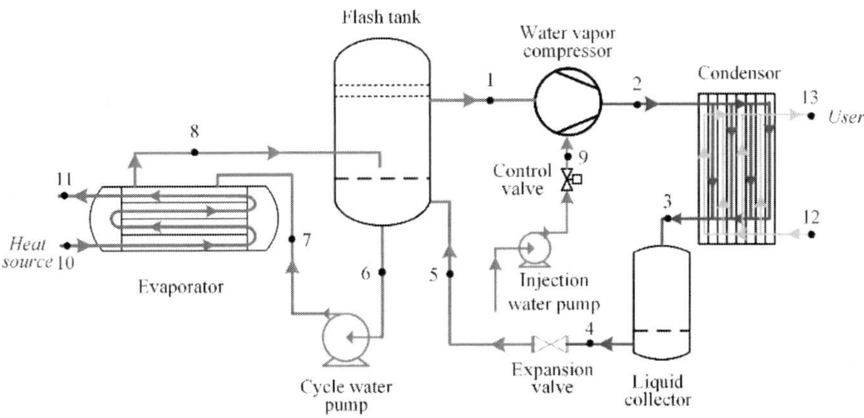


Figure 8. Hot water system by water vapor compression heat pump [84].

Hervás-Blasco et al. [82] utilized the vapor compression heat pump to produce hot water of 60 °C by recovering the heat from the sewage, the schematic is shown in Figure 7. The system used a liquid receiver after the evaporator to ensure the refrigerant at the zero-superheat state and applied subcooling control by the expansion valve to heat the tap water of 10 °C.

The other heat pump is water vapor compression heat pump. Gu et al. [83] adopted it to heat the crude oil to the temperature of 85 °C for transportation and refining, through recovering the by-product water at 60 ± 5 °C. Wu et al. [84] improved the water vapor compression system for 120 ~ 130 °C hot water supply by utilizing 80 ~ 90 °C waste heat, and the schematic is shown in Figure 8. The flash tank is the key component to generate water vapor for the vapor compressor cycle because water is the working medium of the heat pump. The compression process in the compressor is adiabatic, the injection water pump is used for cooling because the discharging temperature would exceed 250 °C at the design pressure ratio.

Xu et al. [85] added the vapor absorption heat pump in a power plant to the produce hot water of 80 °C by the recover the waste heat from the steam condenser, and the schematic is shown in Figure 8. This waste heat recovery system had three advantages: (a) reducing the steam consumption by using the waste heat, (b) reducing the irreversible loss by increasing the temperature of initial hot water, and (c) reducing the load and water consumption of the cooling water.

## 4.2. Space Heating

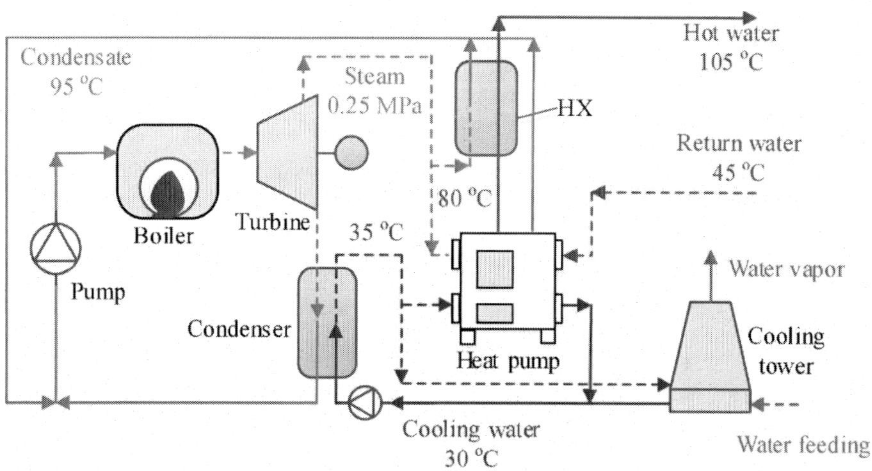


HX: heat exchanger.

Figure 9. Hot water system by vapor absorption heat pump [85].

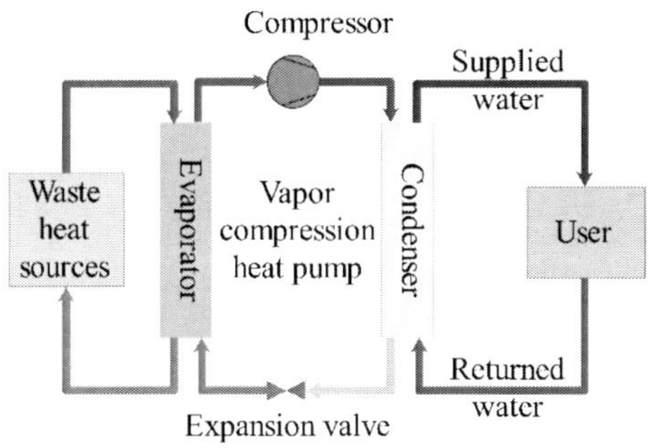


Figure 10. Space heating system by vapor compression heat pump. Modified from [90-93].

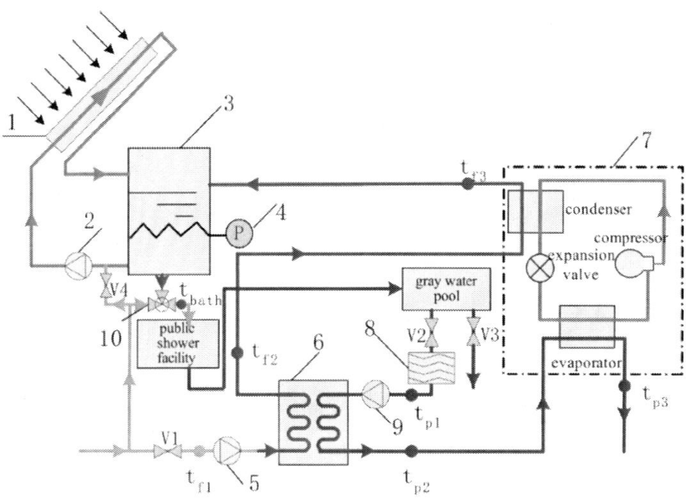


(1) solar collector, (2) pump for collector, (3) heat storage tank, (4) electric heater, (5) water pump, (6) heat exchanger, (7) vapor compression heat pump, (8) filter, (9) waste water pump, (10) tee joint mixing valve.

Figure 11. Space heating system by vapor compression heat pump for greywater recovery [93].

Space heating uses local heat sources to satisfy the heating demand of users by the thermal pipe network [86, 87] and it can be effectively supplied by a heat pump [88]. Heat pump for space heating can recover waste heat sources, such as industrial waste heat, water treatment plant [89], data center air [90], untreated urban sewage [91], and municipal controlled landfills [92], amongst others.

Vapor compression heat pump is widely used for space heating, and the schematic is shown in Figure 10. Deymi-Dashtebayaz et al. [90] utilized the data center as the heat source to space heating for the adjacent office building with an area of 416 $m^2$, because the data center produced a considerable amount of waste heat, which can be recovered by air source heat pump. Liu et al. [91] designed an untreated sewage source heat pump to recover the sewage of 13 °C for space heating, and the key facility of system auto-avoiding-clogging equipment to continuously capture suspended solids in the sewage, resolving the problems of filtration and fouling for the heat exchanger. Liu et al. [93] designed a vapor compression heat pump heating system which is assisted by solar energy and electric heater, and the schematic is shown in Figure 11. The heat pump system can recover the heat from the grey water and utilize solar energy for space heating for hotels and buildings.

Vapor absorption heat pump can also be applied for space heating, as it is shown in Figure 12. The waste heat goes into the generator of the heat pump, the heat exchanger, and the evaporator of the heat pump to supply thermal energy, then the absorber and condenser of the heat pump and heat exchanger absorb the energy to heat the returned water for the user [94].

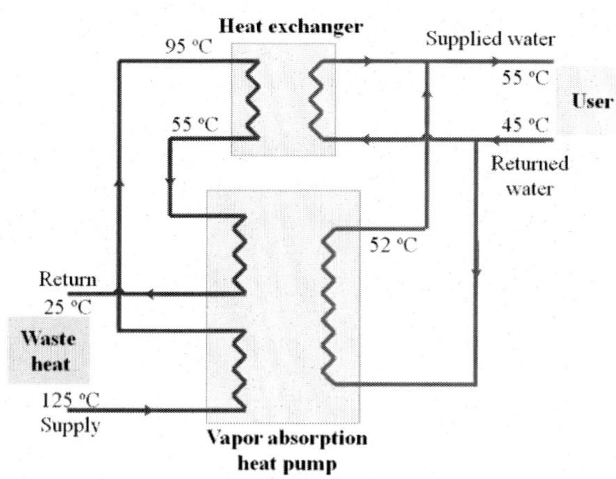


Figure 12. Space heating system by vapor absorption heat pump. Modified from [94].

## 4.3. Snow Melting

Snow melting system can prevent traffic accident and keep people safe [95, 96], and it has been developed as several alternative sustainable systems such as the electrical melting system [97, 98], heat pipe melting system [99, 100], phase-change melting system [101, 102], and hydronic melting system [103, 104]. Among the four novel snow removal techniques, the hydronic melting system is the most practical application because: (a) its heat supply is more reliable than heat pipe and phase-change melting systems, (b) its heat source is more flexible than electrical melting and heat pipe system, and (c) its thermal efficiency is higher than electrical melting system, which can adopt heat pump to drive the hydronic system.

The temperatures of some waste heat are slightly higher than the ambient, such as the water treatment plant [89], data center air [90] and untreated urban sewage [91]. It is important to adopt the waste heat for snow melting to improve energy efficiency and reduce initial cost. The schematic of the snow melting system is shown in Figure 13, the heat pump extracts heat from the waste heat, and then releases thermal energy to the road by the embedded pipes when it is snowing. Liu et al. [105] analyzed the feasibility of heat pump snow melting systems for bridge decks in six major cities of Canada according to the geological conditions and heating demands. The vapor absorption heat pump was not suitable for the snow melting system because the weather and road surface conditions were time-varying [106], requiring that the heat pump should be adjusted flexibly.

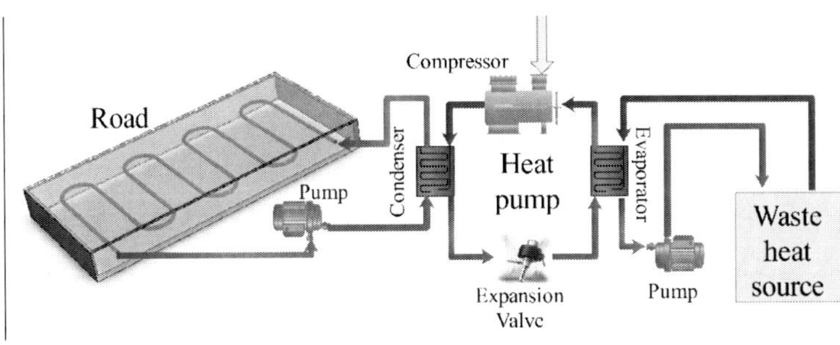


Figure 13. Snow melting system by vapor compression heat pump. Modified from [105].

## 4.4. Conclusion

Water heated by heat pump can be used for hot water supply, space heating, and snow melting in the fields of food, chemical, transportation, residential and commercial buildings, with a water temperature of 40 ~ 120 °C [80]. The types of heat pump applied for hot water production are vapor compression heat pump, vapor absorption heat pump, and solar-assisted vapor compression heat pump.

# 5. WASTE HEAT USED BY HEAT PUMP FOR STEAM PRODUCTION

## 5.1. Sterilization

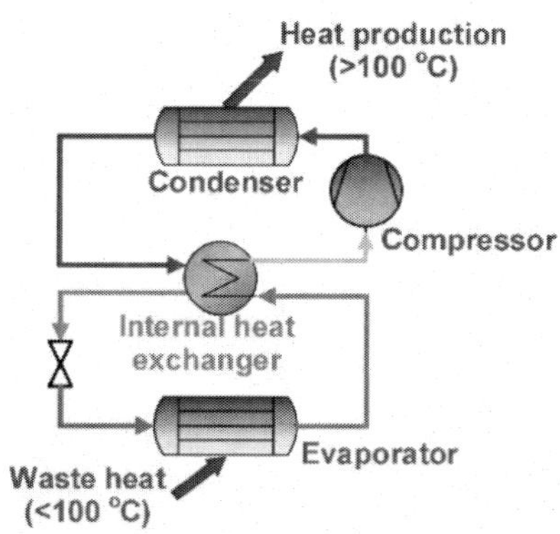


Figure 14. A typical high-temperature heat pump with an internal heat exchanger [80].

Sterilization refers to the process of eliminating microbiological organisms to achieve a sterile environment. Steam sterilization is widely applied in medical equipment [107], food processing, and pharmaceutical industries [108, 109]. However, steam sterilization consumes a large amount of thermal energy [110]. The alternative technology is the high-temperature heat pump for waste heat exploitation to produce steam [80]. Figure 14 shows a typical high-temperature heat pump with an internal heat exchanger, and the internal heat exchanger is mainly used for improving the performance of the heat pump and increasing the temperature of the heat sink (even 150 ~ 160 °C) [111].

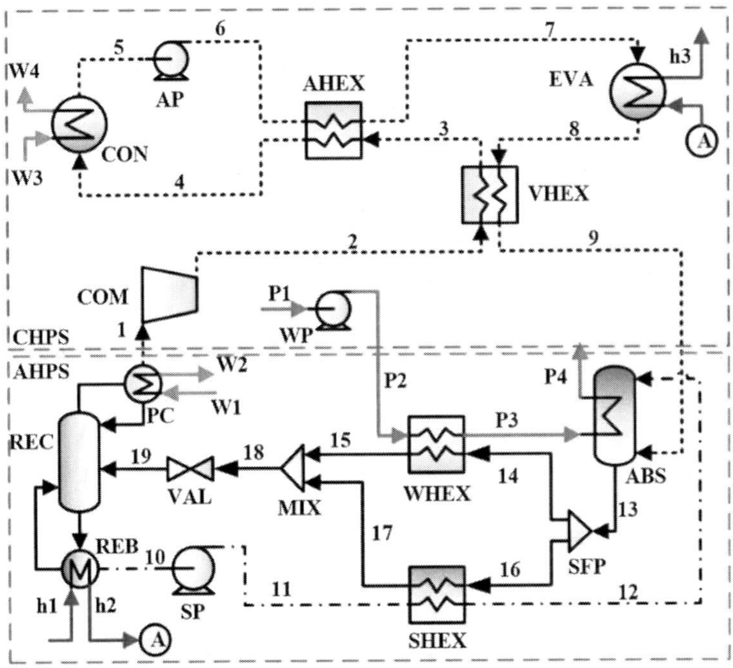


CHPS represents the compression heat pump subsystem, where (1) CON: condenser, (2) AP: ammonia pump, (3) AHEX: ammonia heat exchanger, (4) EVA: evaporator, (5) VHEX: vapor heat exchanger, (6) COM: compressor, (7) WP: water pump.

AHPS represents the absorption heat pump subsystem, where (1) PC: partial condenser, (2) REC: rectifier, (3) REB: reboiler, (4) VAL: valve, (5) SP: solution pump, (6) MIX: mixer, (7) WHEX: water heat exchanger, (8) SHEX: solution heat exchanger, (9) SFP: steam splitter, (10) ABS: absorber.

Figure 15. Schematic of the ammonia-water absorption-compression heat pump [112].

Liu et al. [112] designed an absorption-compression heat pump (as shown in Figure 15) to produce saturated steam at 0.5 MPa (151.8 °C) by utilizing the flue gas below 150 °C as the heat source. The flue gas from a boiler or a heat recovery steam generator was led into the reboiler of the absorption heat pump subsystem and evaporator of the compression heat pump subsystem in sequence to make full use of the waste heat. The green line from P1 to P4 is the forming process of the saturated steam: the ambient water was pressurized by the water pump to 0.5 MPa, then it was preheated by the water heat exchanger to the subcooled stage, and eventually became saturated steam in the absorber.

## 5.2. Distillation

Steam distillation is a special type of distillation to extract and purify the temperature-sensitive materials, such as the natural aromatic compounds and essential oils. The steam can reduce the distillation temperature and residence time, preventing the decomposition or polymerization due to the torrential characteristics of steam [113, 114].

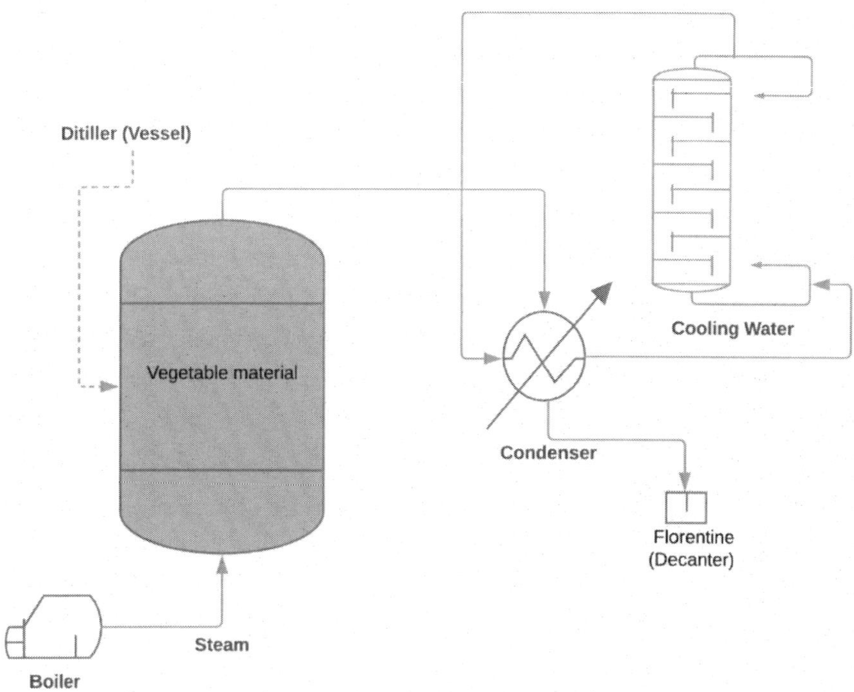


Figure 16. Schematic of the steam distillation for essential oil extraction [115].

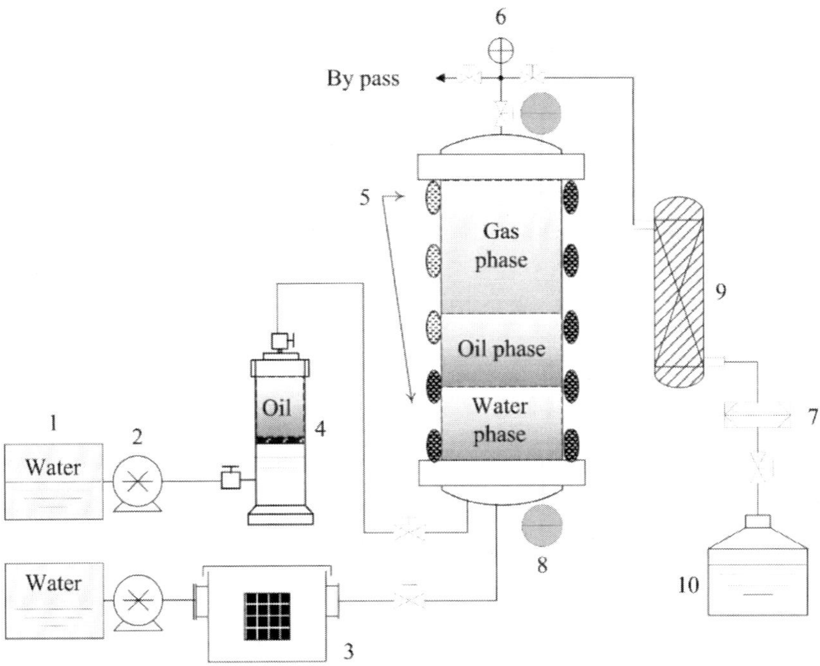


(1) Water tank, (2) Pump, (3) Steam generator, (4) Piston container, (5) Heater bands, (6) Pressure gauge, (7) Back pressure value, (8) Distillation kettle, (9) Condenser pipe, (10) Collection bottle.

Figure 17. Schematic of the steam distillation for heavy oil recovery [116].

Valderrama et al. [115] designed a steam distiller to extract essential oils (as shown in Figure 16), and the boiler can be replaced by the heat pump to produce steam. The steam was transported to the vessel and extracted the molecules of essential oil from the plant. Then the mixture went to the distiller and reached the condenser, and the oil and water were separated in the decanter. Liu et al. [116] utilized a steam distillation apparatus to recover the heavy oil (as shown in Figure 17), and the steam generator was a high-temperature heat pump. The distillations were removed from the top of the distillation kettle, and then they were condensed and collected by a condenser and a collection, respectively.

## 5.3. Rectification

Steam rectification is defined as the multi-stage saturated steam distillation [117], and it is a very efficient technique for gas-liquid separation, which is widely applied in the chemical, petrochemical engineering, and so on [118]. Steam rectification can be operated at a lower temperature than the rectification of binary organic systems according to the Gibbs phase rule [119]. The main difference between steam rectification and steam distillation is the mixture column for steam and original material. Therefore, the steam rectification system can also use the steam produced by the heat pump.

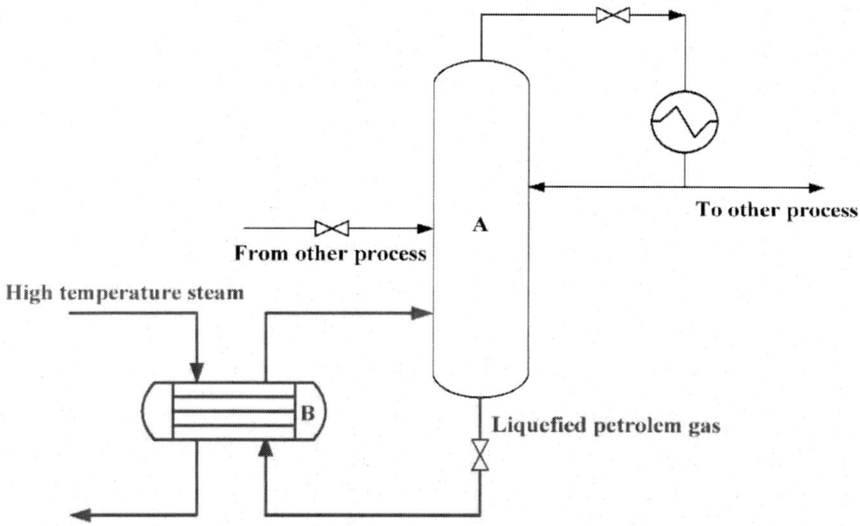


A: Depropane column, B: Bottom reboiler of depropane column.

Figure 18. Schematic of the steam rectification for gas separation [120].

The liquefied petroleum gas is obtained by catalytic cracking and delayed coking units, and it can be separated into propylene ($C_3H_6$), propane ($C_3H_8$), and a mixture of $C_4H_X$ through a series of gas separation units. Li et al. [120] combined a high-temperature heat pump with a gas separation unit to recover waste heat and replace low-pressure steam (120 ~ 130 °C) in the conventional refinery process, and the schematic is shown in Figure 18. The depropane column can be regarded as a rectification system, and the mixture of propane and propylene are separated from the top of the column, whereas the $C_4H_X$ mixture is separated from the bottom.

### 5.4. Conclusion

Steam produced by the heat pump can be used for sterilization, distillation, and rectification in the fields of food, paper, textiles, non-metallic minerals, non-ferrous metal, steel, chemical, and wood products, with the steam temperature of 110 ~ 200 °C [80]. The types of heat pump applied for steam production are vapor compression heat pump and hybrid absorption-compression heat pump.

## Conclusion and Perspectives

Waste heat utilized by heat pump is mainly the low-temperature waste heat and some middle-temperature waste heat because the current highest temperature of heat source used by the heat pump is around 120 °C [111].

Waste heat used by heat pump can be applied for gas heating (drying, preheating, electricity generation, etc.), hot water production (hot water supply, space heating, snow melting, etc.), and steam production (sterilization, distillation, rectification, etc.). The application fields include food, paper, textiles, non-metallic minerals, non-ferrous metal, steel, chemical, wood products, transportation, residential and commercial buildings, and so on. The main types of heat pumps are vapor compression heat pump, vapor absorption heat pump, vapor compression heat pump with organic Rankine cycle, solar-assisted vapor compression heat pump, and hybrid absorption-compression heat pump. The heat pump applications for waste heat recovery are summarized in Table 5.

**Table 5. Heat pump applications for waste heat recovery. Modified from [80]**

| Heat carrier | Typical temperature range (°C) | Application | Heat pump type | Field |
|---|---|---|---|---|
| Gas | 30 ~ 150 | Drying, preheating, and electricity generation | Vapor compression heat pump, vapor absorption heat pump, and vapor compression heat pump with organic Rankine cycle | Food, paper, textiles, non-metallic minerals, non-ferrous metal, steel, and wood products |
| Water | 40 ~ 130 | Hot water supply, space heating, and snow melting | Vapor compression heat pump, vapor absorption heat pump, and solar-assisted vapor compression heat pump | Food, chemical, transportation, residential and commercial buildings |
| Steam | 110 ~ 200 | Sterilization, distillation, and rectification | Vapor compression heat pump and hybrid absorption-compression heat pump | Food, paper, textiles, non-metallic minerals, non-ferrous metal, steel, chemical, and wood products |

The applications of heat pumps for waste heat recovery can effectively narrow the energy gap and reduce negative environmental impacts. However, there are still some barriers for a wider spread of current heat pump technologies for utilizi4ng the waste heat, and the barriers mainly include (1) lack of available refrigerants in the high-temperature range, (2) large installation area, high investment and operational cost, and (3) improper matches between the waste heat source and user side in time, space and energy grade [10, 111].

On the other hand, these barriers challenge and create the future research work. Some perspectives are as follows.

1. Develop environmentally friendly synthetic refrigerants with low GWP (Global Warming Potential) and use the natural refrigerants, such as hydrocarbons, water, and ammonia to find high-temperature refrigerants [111].
2. Increase the heat pump capacity or adopt concentrated waste heat recovery and supply, increase cycle efficiency, and optimize the

components of heat pump and systems to reduce the installation area and cost [10, 111].

3. Use thermal storage, thermal transportation, system optimal operation, continuous turndown ability, and fast response to changes in the operating conditions by using thermodynamic models and control algorithm development to improve the improper matches [10, 18].

## ACKNOWLEDGMENTS

This work was financially supported by the Special Foundation for Major Program of Civil Aviation Administration of China (Grant No. MB20140066), National Natural Science Foundation of China (Grant No. 51976043, 51606050), Heilongjiang Postdoctoral Fund (Grant No. LBH-Z16057), Natural Science Foundation of Heilongjiang Province (Grant No. E2017038), and the Fundamental Research Funds for the Central Universities (Grant No. HIT.NSRIF.2019062).

## REFERENCES

[1] Jouhara H, Khordehgah N, Almahmoud S, Delpech B, Chauhan A, Tassou SA. Waste heat recovery technologies and applications. *Therm Sci Eng Prog* 2018; 6: 268-289.

[2] Firth A, Zhang B, Yang A. Quantification of global waste heat and its environmental effects. *Appl Energy* 2019; 235: 1314-1334.

[3] Forman C, Muritala IK, Pardemann R, Meyer B. Estimating the global waste heat potential. *Renew Sustain Energy Rev* 2016; 57: 1568-1579.

[4] Cullen JM, Allwood JM. Theoretical efficiency limits for energy conversion devices. *Energy* 2010; 35: 2059-69.

[5] Pelda J, Holler S. Spatial distribution of the theoretical potential of waste heat from sewage: A statistical approach. *Energy* 2019; 180: 751-762.

[6] Johnson I, Choate WT, Davidson A. Waste heat recovery: Technology and opportunities in US industry. Laurel, MD (United States): *BCS, Inc.*, 2008.

[7] Advanced Manufacturing Office. Use feedwater economizers for waste heat recovery. *Energy Efficiency and Renewable Energy*, Washington, D.C., 2012.

[8] Yodrak L, Rittidech S, Poomsa-ad N, Meena P. Waste heat recovery by heat pipe air-preheater to energy thrift from the furnace in a hot forging process. *Am J Appl Sci* 2010; 7: 675-681.

[9] Sreejith K, Varghese B, Das D, Devassy D, Harikrishnan K, Sharath GK. Design and cost optimization of plate heat exchanger. *Res Inventy: Int J Eng Sci* 2014; 4: 43-48.

[10] Xu Z, Wang R, Yang C. Perspectives for low-temperature waste heat recovery. *Energy* 2019; 176: 1037-1043.

[11] Tchanche BF, Lambrinos G, Frangoudakis A, Papadakis G. Low-grade heat conversion into power using organic Rankine cycles – A review of various applications. *Renew Sustain Energy Rev* 2011; 15: 3963-3979.

[12] Ogriseck S. Integration of Kalina cycle in a combined heat and power plant, a case study. *Appl Therm Eng* 2009; 29: 2843-2848.

[13] Wang R, Xu Z, Hu B, Du S, Pan Q, Jiang L, Wang LW. Heat pumps for efficient low grade heat uses: from concept to application. *Therm Sci Eng* 2019; 27: 1-15.

[14] Shen C, Lei Z, Wang Y, Zhang C, Yao Y. A review on the current research and application of wastewater source heat pumps in China. *Therm Sci Eng Prog* 2018; 6: 140-156.

[15] Sayegh MA, Jadwiszczak P, Axcell BP, Niemierka E, Bryś K, Jouhara H. Heat pump placement, connection and operational modes in European district heating. *Energy Build* 2018; 166: 122-144.

[16] van de Bor DM, Ferreira CAI. Quick selection of industrial heat pump types including the impact of thermodynamic losses. *Energy* 2013; 53: 312-322.

[17] Chamoun M, Rulliere R, Haberschill P, Peureux JL. Experimental and numerical investigations of a new high temperature heat pump for industrial heat recovery using water as refrigerant. *Int J Refrig* 2014; 44: 177-188.

[18] Goyal A, Staedter MA, Garimella S. A review of control methodologies for vapor compression and absorption heat pumps. *Int J Refrig* 2019; 97: 1-20.

[19] Zimny J, Michalak P, Szczotka K. Polish heat pump market between 2000 and 2013: European background, current state and development prospects. *Renew Sustain Energy Rev* 2015; 48: 791-812.

[20] Blackman C, Gluesenkamp KR, Malhotra M, Yang Z. Study of optimal sizing for residential sorption heat pump system. *Appl Therm Eng* 2019; 150: 421-432.

[21] James NA, Braun JE, Groll EA. The chemical looping heat pump: Thermodynamic modeling. *Int J Refrig* 2019; 98: 302-310.

[22] Bukshaisha AA, Fronk BM. Simulation of membrane heat pump system performance for space Cooling. *Int J Refrig* 2019; 99: 371-381.

[23] Qv D, Wang J, Shang R, Ni L, Yao Y. High-grade energy compensation ratio in a conventional subcritical vapour-compression heat-pump cycle. *Appl Therm Eng* 2018; 138: 490-493.

[24] Li S, Gong G, Peng J. Dynamic coupling method between air-source heat pumps and buildings in China's hot-summer/cold-winter zone. *Appl Energy* 2019; 254: 113664.

[25] Wang G, Wang W, Luo J, Zhang Y. Assessment of three types of shallow geothermal resources and ground-source heat-pump applications in provincial capitals in the Yangtze River Basin, China. *Renew Sustain Energy Rev* 2019; 111: 392-421.

[26] Zhou C, Ni L, Ke Y, Yao Y. Experimental study on the thermal performance of multi-row helically coiled tube heat exchanger for

surface water-source heat pump. *Appl Therm Eng* 2019; 149: 1274-1286.

[27] Jin L, Cao F, Yang D, Wang X. Performance investigations of an R404A air-source heat pump with an internal heat exchanger for residential heating in northern China. *Int J Refrig* 2016; 67: 239-248.

[28] Luo B. Theoretical assessment of an ejector enhanced oil flooded compression cycle. *Int J Refrig* 2017; 73: 154-162.

[29] Luo B, Zou P. Performance analysis of different single stage advanced vapor compression cycles and refrigerants for high temperature heat pumps. *Int J Refrig* 2019; 104: 246-258.

[30] Srikhirin P, Aphornratana S, Chungpaibulpatana S. A review of absorption refrigeration technologies. *Renew Sustain Energy Rev* 2001; 5: 343-372.

[31] Sun J, Fu L, Zhang S, Hou W. A mathematical model with experiments of single effect absorption heat pump using LiBr-$H_2O$. *Appl Therm Eng* 2010; 30: 2753-2762.

[32] Chorak A, Palenzuela P, Alarcón-Padilla DC, Abdellah AB. Energetic evaluation of a double-effect LiBr-$H_2O$ absorption heat pump coupled to a multi-effect distillation plant at nominal and off-design conditions. *Appl Therm Eng* 2018; 142: 543-554.

[33] Zhang J, Zhang H, He Y, Tao W. A comprehensive review on advances and applications of industrial heat pumps based on the practices in China. *Appl Energy* 2016; 178: 800-825.

[34] Vaishak S, Bhale PV. Photovoltaic/thermal-solar assisted heat pump system: Current status and future prospects. *Sol Energy* 2019; 189: 268-284.

[35] Mateu-Royo C, Mota-Babiloni A, Navarro-EsbríJ, Peris B, Molés F, Amat-Albuixech M. Multi-objective optimization of a novel reversible High-Temperature Heat Pump-Organic Rankine Cycle (HTHP-ORC) for industrial low-grade waste heat recovery. *Energy Convers Manage* 2019; 197: 111908.

[36] Urbanucci L, Testi D. Integration of reversible absorption heat pumps in cogeneration systems: Exergy and economic assessment. *Energy Convers Manage* 2019; 200: 112062.

[37] Liu C, Wang Z, Han W, Kang Q, Liu M. Working domains of a hybrid absorption-compression heat pump for industrial applications. *Energy Convers Manage* 2019; 195: 226-235.

[38] International standard industrial classification of all economic activities. Series M, No./Rev. 4. Department of economic and social affairs, statistics division. *United Nations*; 2008.

[39] EPA (Environmental Protection Agency of United States). Waste heat to power systems. *Combined Heat and Power Partnership*, 2012.

[40] Nimbalkar S, Thekdi A. Technologies and materials for recovering waste heat in harsh environments. *Oak Ridge National Laboratory*; 2014.

[41] Department of Energy of United States. *Quadrennial Technology Review: An assessment of energy technologies and research opportunities*, 2015.

[42] Tangjitsitcharoen S, Ratanakuakangwan S, Khonmeak M, Fuangworawong N. Investigation of regenerative and recuperative burners for different sizes of reheating furnaces. *Int J Mech Aerospace Ind Mechatronic Manuf Eng* 2013; 7: 2027-2031.

[43] Dobrego KV, Gnesdilov NN, Kozlov IM, Bubnovich VI, Gonzalez HA. Numerical investigation of the new regenerator–recuperator scheme of VOC oxidizer. *Int J Heat Mass Transfer* 2005; 48: 4695-4703.

[44] Dolz V, Novella R, García A, Sánchez J. HD diesel engine equipped with a bottoming Rankine cycle as a waste heat recovery system. Part 1: study and analysis of the waste heat energy. *Appl Therm Eng* 2012; 36: 269-78.

[45] Larsen U, Nguyen TV, Knudsen T, Haglind F. System analysis and haracteriza of a Kalina split-cycle for waste heat recovery on large marine diesel engines. *Energy* 2014; 64: 484-494.

[46] Liu X, Gong G, Wu Y, Li H. Thermal performance analysis of Brayton cycle with waste heat recovery boiler for diesel engines of offshore oil production facilities. *Appl Therm Eng* 2016; 107: 320-328.

[47] LaGrandeur J, Crane D, Eder A. Vehicle fuel economy improvement through thermoelectric waste heat recovery. *DEER Conference*, 2005.

[48] Hendricks TJ, Karri NK, Hogan TP, Cauchy CJ. New perspectives in thermoelectric energy recovery system design optimization. *J Electron Mater* 2013; 42: 1725-1736.

[49] Vance D, Nimbalkar S, Thekdi A, Armstrong K, Wenning T, Cresko J, Jin M. Estimation of and barriers to waste heat recovery from harsh environments in industrial processes. *J Clean Prod* 2019; 222: 539-549.

[50] Weng Y. Conversion process and utilization of low grade heat energy – Power generation and refrigeration using organic working fluid. *Shanghai Jiao Tong University Press*, Shanghai, 2014; p. 9-10.

[51] Thekdi A, Belt C. Waste heat reduction and recovery options for metals industry. In: *Energy technology*. John Wiley & Sons, Inc., 2011; p. 17-24.

[52] Mamun MAA, Biswas S. Waste heat recovery system by using an organic rankine cycle (ORC). *Int J Sci Eng Res* 2012; 3: 1-4.

[53] Milewski J, Krasucki J. Comparison of ORC and Kalina cycles for waste heat recovery in the steel industry. *J Power Technol* 2017; 97: 302-307.

[54] Kostowski W, Pajączek K, Pociecha A, Kalina J, Niedzielski P, Przybył A. Methods of waste heat recovery – A compressor station case study. *Energy Convers Manage* 2019; 197: 11837.

[55] Tian E, He Y, Tao W. Research on a new type waste heat recovery gravity heat pipe exchanger. *Appl Energy* 2017; 188:586-594.

[56] Palomba V, Brancato V, Frazzica A. Thermal performance of a latent thermal energy storage for exploitation of renewables and waste heat: An experimental investigation based on an asymmetric plate heat exchanger. *Energy Convers Manage* 2019; 200: 112121.

[57] Shen C, Wang M, Zhang C, Ni L, Jiang Y. Experimental investigation on a novel sewage-resource-based system with functions of heat recovery and water purification. *Appl Therm Eng* 2019; Journal Pre-proofs.

[58] Kahraman A, Çelebi A. Investigation of the performance of a heat pump using waste water as a heat source. *Energies* 2009; 2: 697-713.
[59] Wang Q, Liu X, Guo X. Application of waste heat recovery technology in union station and analysis of energy efficiency. *Procedia Eng* 2017; 205: 3860-3866.
[60] Gustafsona RD, Hiibelb SR, Childress AE. Membrane distillation driven by intermittent and variable-temperature waste heat: System arrangements for water production and heat storage. *Desalin* 2018; 448: 49-59.
[61] Gingerich DB, Mauter MS. Quantity, quality, and availability of waste heat from United States thermal power generation. *Environ Sci Technol* 2015; 49: 8297-8306.
[62] Ammar Y, Joyce S, Norman R, Wang Y, Roskilly AP. Low grade thermal energy sources and uses from the process industry in the UK. *Appl Energy* 2012; 89: 3-20.
[63] Papapetrou M, Kosmadakis G, Cipollina A, Commare UL, Micale G. Industrial waste heat: estimation of the technically available resource in the EU per industrial sector, temperature level and country. *Appl Therm Eng* 2018; 138 207-216.
[64] Bertrand A, Aggoune R, Maréchal F. In-building waste water heat recovery: An urban-scale method for the characterization of water streams and the assessment of energy savings and costs. *Appl Energy* 2017; 192: 110-125.
[65] Saker D, Vahdati M, Coker P, Millward S. Assessing the benefits of domestic hot fill washing appliances. *Energy Build* 2015; 93: 282-294.
[66] Pakula C, Stamminger R. Electricity and water consumption for laundry washing by washing machine worldwide. *Energ Effi* 2010; 3: 365-382.
[67] Ni L, Lau S, Li H, Zhang T, Stansbury J, Shi J, Neal J. Feasibility study of a localized residential grey water energy-recovery system. *Appl Therm Eng* 2012; 39: 53-62.
[68] Minea V. Drying heat pumps – Part II: Agro-food, biological and wood products. *Int J Refrig* 2013; 36: 659-673.

[69] Chua KJ, Chou SK, Yang WM. Advances in heat pump systems: a review. *Appl Energy* 2010; 87: 3611-3624.

[70] Adapa PK, Sokhansanj S, Schoenau GJ. Performance study of a recirculating cabinet dryer using a household dehumidifier. *Dry Technol* 2002; 20: 1673-1689.

[71] Chua KJ, Chou SK, Ho JC, Hawlader NA. Heat pump drying: recent developments and future trends. *Dry Technol* 2002; 20: 1579-1610.

[72] Taşeri L, Aktaş M, Şevik S, Gülcü M, Seçkin GU, Aktekeli B. Determination of drying kinetics and quality parameters of grape pomace dried with a heat pump dryer. *Food Chem* 2018; 260: 152-159.

[73] Liu S, Li X, Song M, Li H, Sun Z. Experimental investigation on drying performance of an existed enclosed fixed frequency air source heat pump drying system. *Appl Therm Eng* 2018; 130: 735-744.

[74] Cranston J, Askalany A, Santori G. Efficient drying in washer dryers by combining sorption and heat pumping. *Energy* 2019; 183: 683-692.

[75] Yahya M, Fahmi H, Fudholi A, Sopian K. Performance and economic analyses on solar-assisted heat pump fluidized bed dryer integrated with biomass furnace for rice drying. *Sol Energy* 2018; 174: 1058-1067.

[76] Yu H, Gundersen T, Feng X. Process integration of organic Rankine cycle (ORC) and heat pump for low temperature waste heat recovery. *Energy* 2018; 160: 330-340.

[77] Hernandez-Magallanes JA, Heard CL, Best R, Rivera W. Modeling of a new absorption heat pump-transformer used to produce heat and power simultaneously. *Energy* 2018; 165: 112-133.

[78] Tchanche BF, Lambrinos G, Frangoudakis A, Papadakis G. Low-grade heat conversion into power using organic Rankine cycles – A review of various applications. *Renew Sustain Energy Rev* 2011; 15: 3963-3979.

[79] Cortés E, Rivera W. Exergetic and exergoeconomic optimization of a cogeneration pulp and paper mill plant including the use of a heat transformer. *Energy* 2010; 35: 1289-1299.

[80] Kosmadakis G. Estimating the potential of industrial (high-temperature) heat pumps for exploiting waste heat in EU industries. *Appl Therm Eng* 2019; 156: 287-298.

[81] Wang Z, Guo P, Zhang H, Yang W, Mei S. Comprehensive review on the development of SAHP for domestic hot water. *Renew Sustain Energy Rev* 2017; 72: 871-881.

[82] Hervás-Blasco E, Navarro-Peris E, Barceló-Ruescas F, Corberán JM. Improved water to water heat pump design for low-temperature waste heat recovery based on subcooling control. *Int J Refrig* 2019; 106: 374-383.

[83] Gu Z, Qiu J, Li Y, Cai G. Heat pump system utilizing produced water in oil fields. *Appl Therm Eng* 2003; 23: 1959-1970.

[84] Wu D, Yan H, Hu B, Wang R. Modeling and simulation on a water vapor high temperature heat pump system. *Energy* 2019; 168: 1063-1072.

[85] Xu Z, Mao H, Liu D, Wang R. Waste heat recovery of power plant with large scale serial absorption heat pumps. *Energy* 2018; 165: 1097-1105.

[86] Werner S. International review of district heating and cooling. *Energy* 2017; 137: 617-631.

[87] Rhein J, Henze GP, Long N, Fu Y. Development of a topology analysis tool for fifth-generation district heating and cooling networks. *Energy Convers Manage* 2019; 196: 705-716.

[88] Bellos E, Tzivanidis C, Moschos K, Antonopoulos KA. Energetic and financial evaluation of solar assisted heat pump space heating systems. *Energy Convers Manag* 2016; 120: 306-319.

[89] Hu B, Liu H, Wang R, Li H, Zhang Z, Wang S. A high-efficient centrifugal heat pump with industrial waste heat recovery for district heating. *Appl Therm Eng* 2017; 125: 359-365.

[90] Deymi-Dashtebayaz M, Valipour-Namanlo S. Thermoeconomic and environmental feasibility of waste heat recovery of a data center using air source heat pump. *J Clean Prod* 2019; 219: 117-126.

[91] Liu Z, Ma L, Zhang J. Application of a heat pump system using untreated urban sewage as a heat source. *Appl Therm Eng* 2014; 62: 747-757.

[92] Emmi G, Zarrella A, Zuanetti A, Carli MD. Use of municipal solid waste landfill as heat source of heat pump. *Energy Procedia* 2016; 101: 352-359.

[93] Liu L, Fu L, Jiang Y. Application of an exhaust heat recovery system for domestic hot water. *Energy* 2010; 35: 1476-1481.

[94] Fang H, Xia J, Jiang Y. Key issues and solutions in a district heating system using low-grade industrial waste heat. *Energy* 2015; 86: 589-602.

[95] Mensah K, Choi J. Review of technologies for snow melting systems. *J Mech Sci Technol* 2015; 29: 5507-5521.

[96] Han C, Yu X. An innovative energy pile technology to expand the viability of geothermal bridge deck snow melting for different United States regions: Computational assisted feasibility analyses. *Renew Energy* 2018; 123: 417-427.

[97] Sun Y, Wu S, Liu Q, Hu J, Yuan Y, Ye Q. Snow and ice melting properties of self-healing asphalt mixtures with induction heating and microwave heating. *Appl Therm Eng* 2018; 129: 871-883.

[98] Lai Y, Liu Y, Ma D. Automatically melting snow on airport cement concrete pavement with carbon fiber grille. *Cold Reg Sci Technol* 2014; 103: 57-62.

[99] Wang X, Zhu Y, Zhu M, Zhu Y, Fan H, Wang Y. Thermal analysis and optimization of an ice and snow melting system using geothermy by super-long flexible heat pipes. *Appl Therm Eng* 2017; 112: 1353-1363.

[100] Ozsoy A, Yildirim R. Prevention of icing with ground source heat pipe: a theoretical analysis for Turkey's climatic conditions. *Cold Reg Sci Technol* 2016; 125: 65-71.

[101] Cunha JP, Eames P. Thermal energy storage for low and medium temperature applications using phase change materials – a review. *Appl Energy* 2016; 177: 227-238.

[102] Farnam Y, Esmaeeli HS, Zavattieri PD, Haddock J, Weiss J. Incorporating phase change materials in concrete pavement to melt snow and ice. *Cement Concr Compos* 2017; 84: 134-145.

[103] Wang H, Liu L, Chen Z. Experimental investigation of hydronic snow melting process on the inclined pavement. *Cold Reg Sci Technol* 2010; 63: 44-49.

[104] Zhou Z, Wang X, Zhang X, Chen G, Zuo J, Pullen S. Effectiveness of pavement-solar energy system – An experimental study. *Appl Energy* 2015; 138: 1-10.

[105] Liu H, Maghoul P, Bahari A, Kavgic M. Feasibility study of snow melting system for bridge decks using geothermal energy piles integrated with heat pump in Canada. *Renew Energy* 2019; 136: 1266-1280.

[106] Liu X, Ree SJ, Spitler JD. Modeling snow melting on heated pavement surfaces. Part II: Experimental validation. *Appl Therm Eng* 2007; 27: 1125-1131.

[107] Feurhuber M, Cattide A, Magno M, Miranda M, Prieler R, Hochenauer C. Prediction of the fluid flow, heat transfer and inactivation of microorganism at medical devices in modern steam sterilizers using computational fluid dynamics. *Appl Therm Eng* 2017; 127: 1391-1403.

[108] Lau WL, Reizes J, Timchenko V, Kara S, Kornfeld B. Heat and mass transfer model to predict the operational performance of a steam sterilisation autoclave including products. *Int J Heat Mass Transfer,* 2015; 90: 800-811.

[109] Cooper SJG, Hammond GP, Hewitt N, Norman JB, Tassou SA, Youssef W. Energy saving potential of high temperature heat pumps in the UK Food and Drink sector. *Energy Procedia*, 2019; 161: 142-149.

[110] Simpson R, Abakarov A. Optimal scheduling of canned food plants including simultaneous sterilization. *J Food Eng 2009*; 90: 53-59.

[111] Arpagaus C, Bless F, Uhlmann M, Schiffmann J, Bertsch SS. High temperature heat pumps: market overview, state of the art, research

status, refrigerants, and application potentials. *Energy* 2018; 152: 985-1010.

[112] Liu C, Jiang Y, Han W, Kang Q. A high-temperature hybrid absorption-compression heat pump for waste heat recovery. *Energy Convers Manage* 2018; 172: 391-401.

[113] Chen L, Liu Y, Deng J. Removal of phthalic acid esters from sea buckthorn (*Hippophae rhamnoides* L.) pulp oil by steam distillation and molecular distillation. *Food Chem* 2019; 294: 572-577.

[114] Chemat F, Lucchesi ME, Smadja J, Favretto L, Colnaghi G, Visinoni F. Microwave accelerated steam distillation of essential oil from lavender: a rapid, clean and environmentally friendly approach. *Anal Chim Acta* 2006; 555: 157-160.

[115] Valderrama F, Ruiz F. An optimal control approach to steam distillation of essential oils from aromatic plants. *Comput Chem Eng* 2018; 117: 25-31.

[116] Liu P, Yuan Z, Zhang S, Xu Z, Li X. Experimental study of the steam distillation mechanism during the steam injection process for heavy oil recovery. *J Pet Sci Eng* 2018; 166-561-567.

[117] Zhang L, Xue Z, Ni L, Ma D. Extraction and quality analysis of volatile oils from onions by coupling pilot and laboratory equipment based on multi-rectification. *Sep Purif Technol* 2014; 137: 36-42.

[118] Jiang W. Unit Operation. *Qinghua University Press*, China, Beijing, 2005.

[119] Qian Q, Wang H, Bai P, Yuan G. Effects of water on steam rectification in a packed column. *Chem Eng Res Des* 2011; 89: 2560-2565.

[120] Li X, Zhang Y, Fang L, Jin Z, Zhang Y, Yu X, Ma X, Deng N, Wu Z. Energy, exergy, economic, and environmental analysis of an integrated system of high-temperature heat pump and gas separation unit. *Energy Convers Manage* 2019; 198: 111911.

In: Waste Heat: Sources and Uses
Editor: Michèle Picard

ISBN: 978-1-53617-355-0


*Chapter 2*

# EJECTOR ENHANCED WASTE HEAT RECOVERY SYSTEMS EFFICIENCY

***Zine Aidoun*[1], *Khaled Ameur*[1], *Messaoud Badache*[1] and *Hakim Nesreddine*[2,*]**

[1]Buildings group, CanmeENERGY, Varennes, Canada
[2]Hydro-Québec, Shawinigan, Canada

## ABSTRACT

Energy consumption for industrial, commercial and domestic needs continues to increase and supplies to satisfy this demand with low environmental impact represents a major challenge. At the same time, considerable amounts of energy input to industrial activities are released in the environment at low temperature. Waste Heat Recovery (WHR) is identified among the most appropriate solutions to this issue, however available waste heat literature shows that technologies to extract and use low-grade waste heat effectively are scarce.

In the recent literature devoted to WHR activities, the ejector technology has been receiving particular attention, due to its suitability to harness waste heat at the lower range of temperatures where other established technologies are no longer efficient.

* Corresponding Author's Email: Nesreddine.Hakim@ireq.ca.

The use of ejectors as low temperature waste heat transformers to produce cooling and compression effects has been abundantly discussed in the refrigeration and heat pumping literature. Integrated in WHR systems in order to build various forms of combination with other equipment, this device has gained renewed interest as a solution for low temperature heat recovery to enhance overall efficiency, reduce costs and control greenhouse gases (GHG) emissions. Uses include cooling and heating operations, industrial process enhancement as a heat exchanger or integrated in systems to improve their overall efficiency as a circulating, expander, pumping or compression agent.

In the present chapter, the ejector potential to enhance a variety of conventional technologies for waste heat recovery has thus been the main objective, especially at decreasing waste heat temperature grades. The focus is on the ejector use as a component in processes, in cycles to form hybrid units or integrated in systems as a device to improve their operation and performance. The result is a summary of the main findings and trends, in the area of ejector-enhanced efficiency of WHR cycles and systems using low boiling point working fluids. Following a brief overview presentation of the ejector, selected configurations of interest, are presented and discussed in a variety of application domains according to three main topics: refrigeration-heating systems, power systems and tri-generation systems.

**Keywords**: waste heat recovery, low-grade energy, ejector integration, combined cycles, review

## 1. Introduction

Systems relying on ejectors to increase their performances and reduce their environmental footprints are attracting very much attention, in the context of waste heat recovery technologies. As a thermal compression device activated by low-grade heat (single-phase version) or as a pressurized liquid expander for fluid circulation (two-phase version) the ejector distinctive characteristics are geometric simplicity, low operation/ maintenance costs and absence of moving parts, making it attractive for many applications. Used in the past for pumping gases from systems to produce vacuum or compressing vapors, ejectors have thus been around for a long time but mainly restricted to industrial applications using steam as the working fluid (ASHRAE 1965).

Steam ejectors are commercially available in different sizes and geometries to match different duty requirements; however, for application environments with other fluids and conditions, ejectors require specific designs or are yet at various stages of development. For example, in the case of heat pumps (heat upgrading, cooling and refrigeration), applications are still limited, due to the lack of knowledge on their performance over a wide range of conditions with different refrigerants, and also because of their low energy efficiency compared to mechanical vapor compression systems. However, in the current energy and environmental context, ejectors may become instrumental to provide a reliable and low-cost solution for recovering and upgrading thermal energy, provided accurate design and appropriate working fluid selection.

The use of working fluids other than steam such as organic fluids, natural and new synthetic refrigerants with no impact on the environment open therefore many new doors of application to ejectors (Besagni, Mereu, and Inzoli 2016; Aidoun et al. 2019a; 2019b).

In some refrigeration systems they can play a central role, like substituting for the compressor or the throttling valve, in other cases they are integrated as enhancement agents in industrial processes or they form hybrid systems with other known conventional technologies. Recent reviews provide more details on ejector design and analysis complemented with several examples, mostly in heating and cooling applications (Aidoun et al. 2019a; 2019b; Besagni 2019; Tashtoush, Al-Nimr, and Khasawneh 2019).

In this chapter, some important features of ejectors are highlighted. Update studies of their potential applications are reported in the next paragraphs with particular focus on industrial waste heat recovery technologies integrating ejectors in various configurations enhancing overall efficiency, reducing costs and promoting environmental benefits.

## 2. Main Ejector Types

Several types of ejectors are described in the literature, according to the nature of the flow and the geometry, generally related with operational and physical constraints of the application (Figure 1). For most applications, axisymmetric ejectors are preferred for reasons of easier manufacturing and cost. More details are available on these features in the above-mentioned reviews.

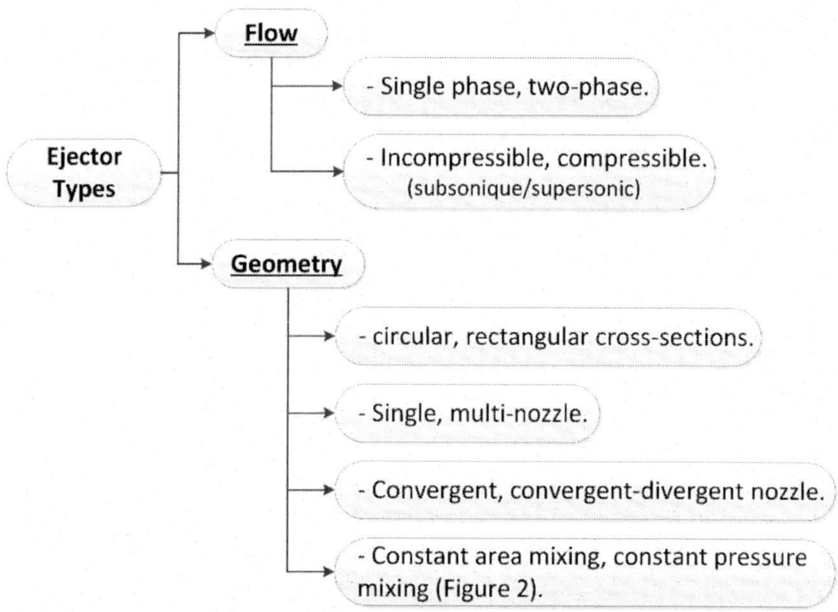


Figure 1. Main ejector types.

## 3. Ejector Operation

An ejector is a passive compression device with no moving parts (Figure 2). Activated by the thermal or potential energy of a primary flow expanding in a convergent-divergent nozzle, the ejector entrains and compresses a secondary flow. Motive and induced flows usually merge at supersonic conditions in a mixing chamber and the resulting mixture, after a series of shocks, is compressed in a diffuser to the outlet pressure. Generally, low-grade recovered waste heat is the energy source in single-phase ejectors and depending on its quality and availability, it may be used to provide a compression duty for refrigeration needs or for tasks of fluid transfer in several industrial and process applications. Ejectors are sensitive to internal geometry, nozzle relative positions, working fluids and working conditions which affect their capacity and performance.

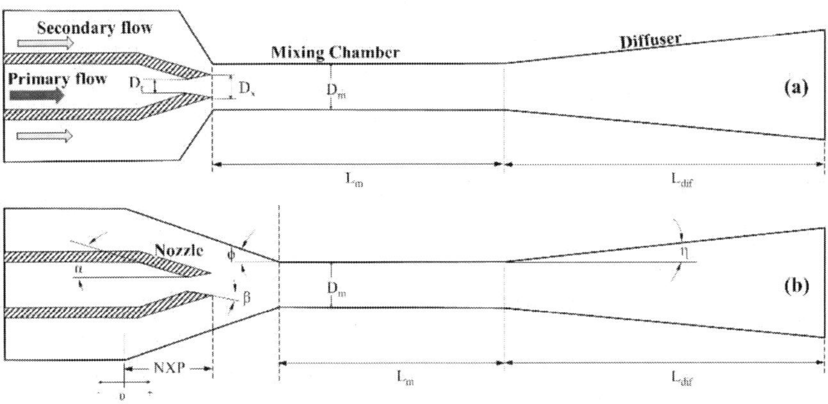


Figure 2. Ejector geometry lay-out (a) constant area mixing ejector and (b) constant pressure mixing ejector.

The two main performance parameters used to qualify ejectors are the entrainment ratio and the compression ratio. The entrainment ratio, ω is defined as the mass flowrate ratio:

$$\omega = \frac{\dot{m}_{secondary}}{\dot{m}_{primary}} \qquad (1)$$

and the compression ratio τ as the pressure ratio:

$$\tau = \frac{P_{back}}{P_{secondary}} \qquad (2)$$

For an ejector with fixed geometry, the performance is evaluated at design conditions. Under these conditions, the secondary and primary flows at their respective nozzle throats, reached sonic conditions. The ejector performance is then maximized.

Figure 3 is the ejector characteristic curve representing the variation of the entrainment ratio, ω as a function of the backpressure (ejector outlet) for a fixed geometry and inlet conditions. Three zones can be distinguished on this figure. Pressure $P_b$* corresponds to the critical condition, generally referred to as the ejector design point for which the primary mass flowrate is maximum and the flow beyond the primary nozzle throat is supersonic. The secondary mass flowrate in the same conditions is also maximum and the flow has reached the sonic velocity. The flow is said in double choking (i.e., on design) condition.

In the zone $0 \leq P \leq P_b^*$, the entrainment ratio ω remains constant, for both mass flowrates are choked. In the zone where $P_b^* \leq P \leq P_{b0}$ the primary flow is choked but the secondary flow is subsonic and varies with backpressure. The ejector works in off design (i.e., sub-critical) mode. Beyond $P_{b0}$ the ejector malfunctions (i.e., reverse flow mode).

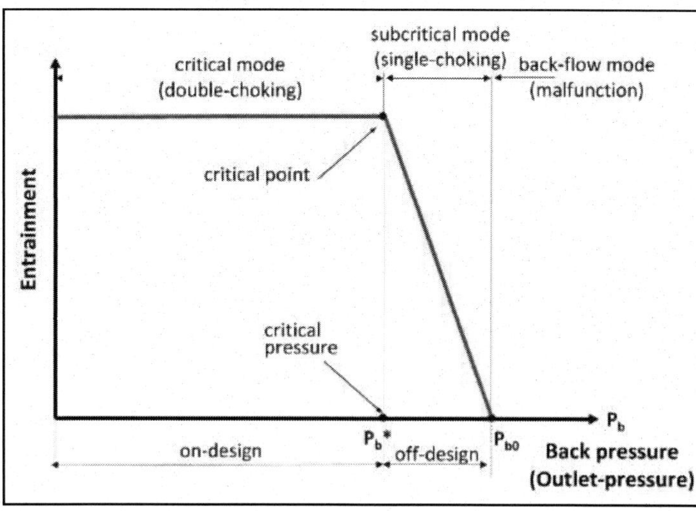


Figure 3. Ejector characteristic curve with operation zones.

It may be inferred from the above discussion that an ejector of fixed geometry has a narrow range of efficient operation around its design point. Beyond this threshold, the ejector falls quickly in off design mode and quickly degrading performance. By modifying internal geometric features, it is possible to restore efficient ejector operation to various extents. Research has demonstrated that several factors diversely influence the ejector performance. The nozzle exit position (NXP) and the mixing throat area ratio $\varphi = (D_m/D_t)^2$ are relevant, but parameters such as the nozzle exit area ratio $\theta = (D_x/D_t)^2$ and the mixing chamber throat length ratio $\delta = L_m/D_m$ do influence performance as well (Aidoun et al. 2019a).

## 4. Potential Areas of Interest for Ejectors

This paragraph intends to show the versatility of ejectors as thermal converters to be combined with a great variety of more complex heat recovery technologies to enhance overall operation efficiency. Representative applications in different areas are highlighted through illustrative research development. An important reminder is that the majority of the studies available in the industrial range are built on thermodynamics, exergoeconomics and considerations of other energy based methods. By combining exergy and conventional concepts of engineering economics they evaluate and optimize the design, performance and cost of energy systems by means of multi-objective procedures. Experimental studies are scarce and generally performed at laboratory scale in the range of few kilowatts capacity (X. Chen et al. 2013; Wang et al. 2016; Abed et al. 2017) and exceptionally at medium scale (Mazzelli and Milazzo 2015).

Two distinctive features make the ejector very attractive in diversified areas of potential applications: the absence of moving parts is a reliability criterion with little or no maintenance and low operational costs. The ability to operate with low and very low-grade heat is a great asset in the area of waste heat and renewable energy harnessing. In addition, ejectors may fulfill multiple roles in processes: as mere components in some industrial instances, as main movers in many heat-activated refrigeration cycles or integrated with other technologies to form more complex energy efficiency systems. This flexibility extends the ejector potential for applications ranging from refrigeration to industrial processes (Figure 4).

Ejector systems operated with low boiling refrigerants can harness solar energy, compressor superheat, heat engine and turbine exhausts of ORC and Kalina cycles by way of example. The technology is common in power and energy fields, food and pharmaceutical processing, chemical, petrochemical, pulp & paper and related industries for mixing fluids of different types, conditions and toxicity (Daubert-Deleris et al. 2006; Fahmy, Mewes, and Ohlrogge 2002). Vacuum creation in food processing included drying, liquid evaporation, juice concentration, gas extraction and odor suppression are among industrial applications (ASHRAE 1983; Akteriana 2011; Reddick et al. 2016; 2018; Bowrey, Dang, and Sergeant 1986; Shaozhi et al. 2018). Other industrial opportunities such as enhancing gas turbine performance with flue gases energizing an ejector to precool inlet air to the feed compressor (Al-ansary 2007; V. Petrenko 2009) or replacing compressors in natural gas recompression stations with ejectors activated by high-pressure gas from the main line to boost low-pressure gas continues to draw attention (Jia Yan et al. 2013). Further applications are also found in aeronautics for engine boosting, thrust augmentation, air fuel mixing and oil radiator cooling (Choutapalli, Anjaneyulu, and Alkislar 2012; Lijo et al. 2012; Shan, Zhang, and Ren 2018) and in many other areas (Aidoun et al. 2019b; 2019a).

Moreover, ejector based refrigeration continues to attract a lot of research. Applications in air conditioning, cooling and heating have been among the priorities of many researchers who proposed very good quality works with a variety of refrigerants (D. W. Sun and Eames 1996; Alexis and Karayiannis 2005; Selvaraju and Mani 2006; Abdulateef et al. 2009; Butrymowicz et al. 2014; Butrymowicz, Śmierciew, and Karwacki 2014; Ruangtrakoon and Aphornratana 2014; Chandra and Ahmed 2014). Car cabin cooling by means of heat recovery from the engine radiator to activate a supersonic ejector has been investigated (L. T. Chen 1978; Salim 2004; Wang et al. 2016; Jaruwongwittaya and Chen 2012).

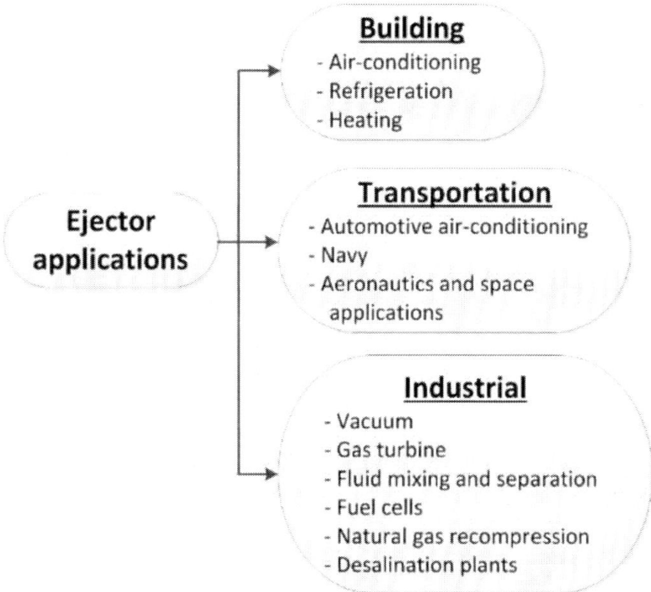


Figure 4. Areas of potential ejector applications.

In the present chapter, the focus will be set on systems powered by waste heat and integrating ejectors for the purpose of cooling, heating or power generation and more generally for energy efficiency enhancements, economic viability and reduced environmental footprint in buildings and industrial plants.

## 5. Working Fluids for Ejectors

Selection criteria of the working fluid include, at least considerations such as environmental preservation, low ODP (Ozone Depletion Potential) and low GWP (Global Warming Potential), safety issues, availability and costs in addition to the primary requirement of performance.

The main families of working fluids used with ejectors are summarized in Figure 5. Early works were performed for validation purposes mostly using air and steam for obvious reasons of simplicity and safety. Tests with conventional refrigerants became subsequently more common, given their predominance in refrigeration. The current tendency towards alternative refrigerants has favored a remarkable increase studies with environmentally friendly natural fluids such as $CO_2$ (V. O. Petrenko, Huang, and Ierin 2011; X. Chen et al. 2013) and hydrocarbons (T. Zhang and Mohamed 2014; Mohamed, Shatilla, and Zhang 2019; Popovac et al. 2017). A significant research effort is also carried out to move to next generation low GWP and non-flammable refrigerants, HFOs and HCFOs (Kasperski and Gil 2014; Fang et al. 2017; Śmierciew et al. 2017; Gagan, Śmierciew, and Butrymowicz 2018; Takleh and Zare 2019).

Ejectors as heat-activated devices are very sensitive to the working fluids in terms of operation and performance. Dry refrigerants (saturation curve skewed right in Figure 6) are preferred to wet ones in this respect because they do not condense when they expand in the primary nozzle. Condensation can be avoided in wet refrigerants by providing limited superheat but this somewhat affects performance. In a refrigeration cycle, and depending on the fluid type, the amount of superheat needs therefore to be maintained within reasonable limits, so as not to waste energy and reduce cycle coefficient of performance (COP).

Favorable thermophysical properties to ensure high performance in the range of intended use is of primary importance. Properties such as latent heat of evaporation, specific heat, viscosity, thermal conductivity, specific volume, normal boiling and critical points all play an important role in ejector operation and overall cycle performance. In order to be able to use low-grade thermal energy, low boiling temperatures are desired as well. In general, a high specific volume increases the ejector size and high molecular weight favors the ejector entrainment ratio but may also reduce size. A high latent heat of evaporation and a wide generator temperature range reduce the specific circulation rate.

Theoretical and experimental investigations for the evaluation of a wide range of refrigerants is available in the literature. Table shows a selection of some relevant studies on this topic. For interested readers, this topic was covered in more detail in previous reviews (Besagni, Mereu, and Inzoli 2016; S. Chen, Chen, and Fang 2015; Abdulateef et al. 2009; Aidoun et al. 2019b; 2019a).

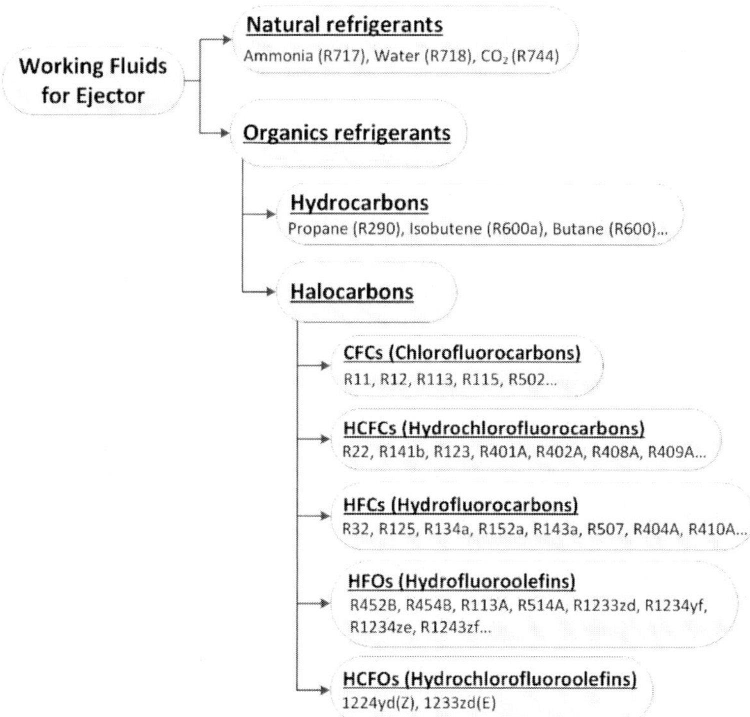


Figure 5. Main fluids used with ejectors.

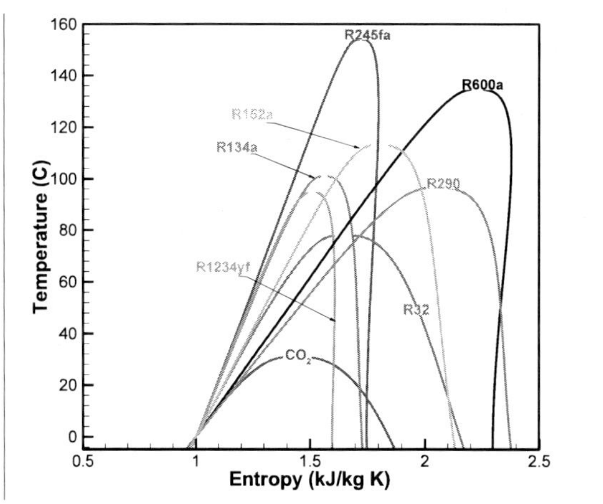


Figure 6. Saturation curves of various refrigerants.

**Table 1. Ejector experiments with steam and organic refrigerants**

| Fluid | Load (kW) | Source conditions | Sink conditions | WH conditions | Reference |
|---|---|---|---|---|---|
| Natural fluids | | | | | |
| Water | 1 | 5-10°C | 26-37°C | 120-140°C | (I.W Eames, Aphornratana, and Haider 1995) |
| | - | 50-90 kPa | 100-170 kPa | 350-550 kPa | (Reddick et al. 2018) |
| | 1 | 7.5°C | 35-70 mbar | 130-140°C | (Kitrattana et al. 2017) |
| $CO_2$ | 23-70 | -30 to -3°C | 35°C | - | (Banasiak et al. 2015) |
| | 46 | -8°C | 26-36°C | - | (Haida et al. 2016) |
| | 29-36 | -5 to 12°C | 40-60°C | - | (Boccardi et al. 2017) |
| | 5 | 50 bar | 35-55°C | - | (Zhu et al. 2018) |
| | 14 | -10.2°C | 33.5°C | - | (Santini et al. 2019) |
| Hydrocarbons | | | | | |
| R600a | 2 | 9°C | 12-30°C | 83°C | (Pereira et al. 2014) |
| | - | 3.5-7°C | 24-34°C | 50-64°C | (Butrymowicz et al. 2014) |
| | 0.3 | 70 kPa | 500 kPa | - | (Jeon et al. 2018) |
| R600 | 20.9 | 53.2°C | 80.4°C | - | (Popovac et al. 2017) |
| R290/R170 | - | -65 to -55°C | 35-45°C | - | (Liu, Yu, and Yan 2018) |
| R601 | - | 8-16°C | 84-95°C | - | (Mohamed, Shatilla, and Zhang 2019) |
| CFCs | | | | | |
| R113 | Up to 2 | 5-15°C | 35-47°C | 65-80°C | (Huang, Jiang, and Hu 1985) |
| | Up to 0.7 | 6.7°C | 42.3°C | 87.8°C | (Al-Khalidy 1997) |
| R11 | Up to 1.7 | -5-12°C | 35-41°C | 100-110°C | (S. Aphornratana, Chungpaibulpatana, and Srikhirin 2001) |
| | ≤ 1 | 0 - 16°C | 27-34°C | 88-102°C | (Yapici and Yetişen 2007) |
| HCFCs | | | | | |
| R142b | 1 | -10 | 30 | 85 | (Hernandez et al. 2004) |
| R123 | - | 10°C | 34°C | 83-103°C | (Yapici et al. 2008) |

| Fluid | Load (kW) | Source conditions | Sink conditions | WH conditions | Reference |
|---|---|---|---|---|---|
| HCFCs | | | | | |
| R141b | 2 | -6-15°C | 30-33°C | 70-130°C | (Thongtip and Aphornratana 2015) |
| | 2 | -6-20°C | 24-36°C | 70-140°C | (Thongtip and Aphornratana 2017) |
| HFCs | | | | | |
| R236fa | 1.5 | 4-12°C | 27-33°C | 60-100°C | (I.W. Eames, Petrenko, and Ablwaifa 2004) |
| R245fa | 2 | 8-15°C | 34-44°C | 100-120°C | (Ablwaifa, Eames, and Petrenko 2009) |
| | 12 | 8-16°C | 30-39°C | 80-105°C | (Shestopalov et al. 2015) |
| | 35 | 10-20.5°C | 33-41°C | 74-90°C | (Hamzaoui et al. 2018) |
| R134a | 0.5 | 2-13.5°C | 26-38°C | 65-90°C | (Selvaraju and Mani 2006) |
| | - | 1.5-24°C | 12-24°C | 72-90°C | (Jiwei Yan et al. 2017) |
| R410A | 1 | 4-9°C | 35°C | - | (Lawrence and Elbel 2016) |
| | 12 | 8-20°C | 41-51°C | - | (Kim et al. 2018) |
| R404A | - | -10 to 0°C | 50°C | - | (Man-O et al. 2007) |
| R134a/R23 | 0.1 | -50 to -40°C | 15-27°C | - | (Bai, Yan, and Yu 2018) |
| HFOs | | | | | |
| R1234yf | - | -10 to 10°C | 30-55°C | - | (H. Li et al. 2014) |
| R1234ze(E) | 30 | -6 to 13°C | 18-30°C | 53-70°C | (Śmierciew et al. 2017) |
| HCFOs | | | | | |
| R1224yd(Z) R1233zd(E) | - | 0-12 | 25-40 | 78-95 | (Gil and Kasperski 2018) |

# 6. Ejectors in Refrigeration and Heating Systems

## 6.1. Ejector Refrigeration Systems (ERS)

A simple ejector set-up ERS is depicted in Figure 7. It has been generally used in the literature as a base reference for comparisons and/or as a building block for more complex cycles (Selvaraju and Mani 2006; García Del Valle et al. 2014; Jiwei Yan et al. 2017; S. He, Li, and Wang 2009). In the context of cold production, the thermal coefficient of performance is defined as:

$$COP_{th} = \omega \frac{Q_E}{Q_G + W_p} \quad (3)$$

A mechanical formulation is expressed as:

$$COP_{me} = \omega \frac{Q_E}{W_p} \quad (4)$$

Where $Q_G$, $Q_E$ and $W_p$ are specific heats and mechanical work transferred during the process.

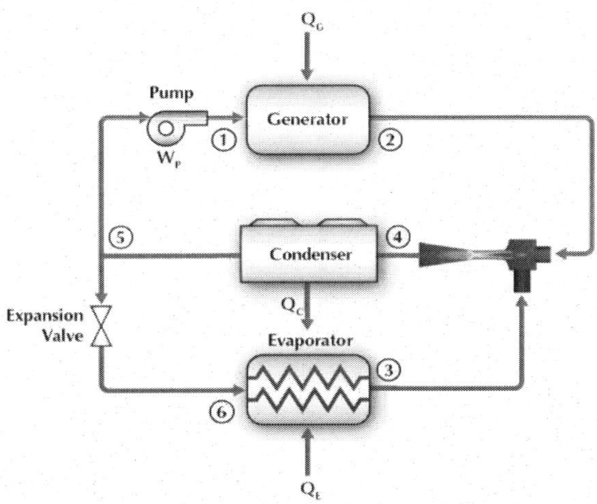


Figure 7. Schematic representation of a conventional ERS.

An example of a compound cycle built on this principle was proposed by G. Chen et al. (2017). The authors carried out experiments with R236fa on a two-stage ejector cycle driven by two heat sources at different temperatures (Figure 8). A high temperature source (High-T generator) drove the low-P ejector connected to the evaporator, whose outlet flow was the secondary stream of a high-P ejector. Outlet flow from this latter was then condensed and split into a first stream returning to the evaporator and a second stream pumped back to the conditions of the High-T generator. This stream was further split into the portion feeding the High-T generator and the one expanded to the conditions of the low-T generator to activate the High-P ejector. This system was tested over a wide range of conditions affecting its operation and performance. Results indicated that this system could supply a superior refrigerating effect than a single-stage conventional ERS at condensing temperatures lower than 21.7°C, when all other conditions were similar.

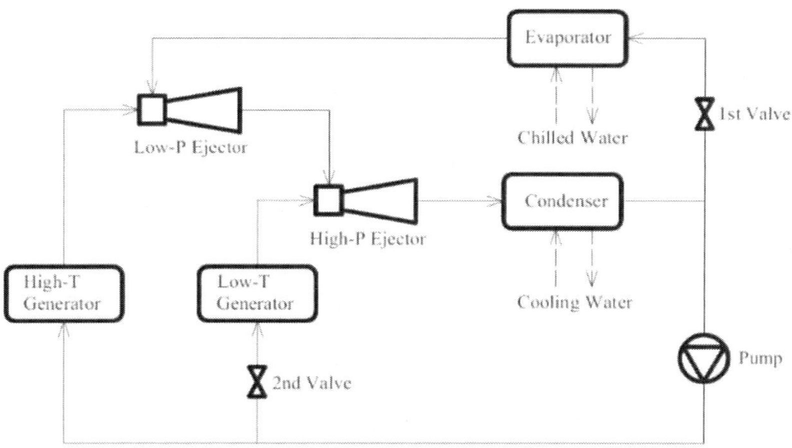


Figure 8. Two-stage ejector cycle (G. Chen et al. 2017).

A related configuration in the form of ERS-compression cascade prototype was tested by Nesreddine et al. (2015) for low temperature operations. The system combines a sub-critical $CO_2$ vapor compression loop serving as low stage with a high stage ejector compression loop working on R245fa which may be powered by any free or low-cost heat source (Figure 9). This system provided a refrigeration capacity of 15 kW at -20°C and was thermally activated by low-grade heat. Tests were performed within the following parameters ranges: $25°C \leq T_{sink} \leq 35°C$, $-25°C \leq T_{source} \leq -15°C$, $T_{waste\ heat}=80°C$. Experiments showed that the cascade system COP for $T_{source}=-20°C$ and $T_{sink}=25°C$ was superior to a single stage $CO_2$ compression system by over 40%.

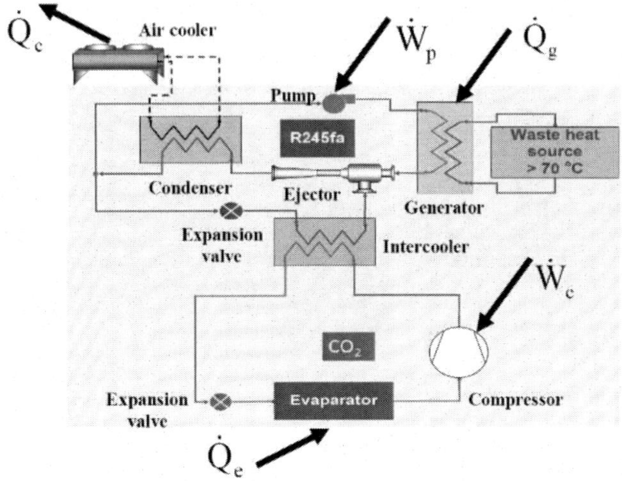


Figure 9. Cascade ejector compression system (Nesreddine et al. 2015).

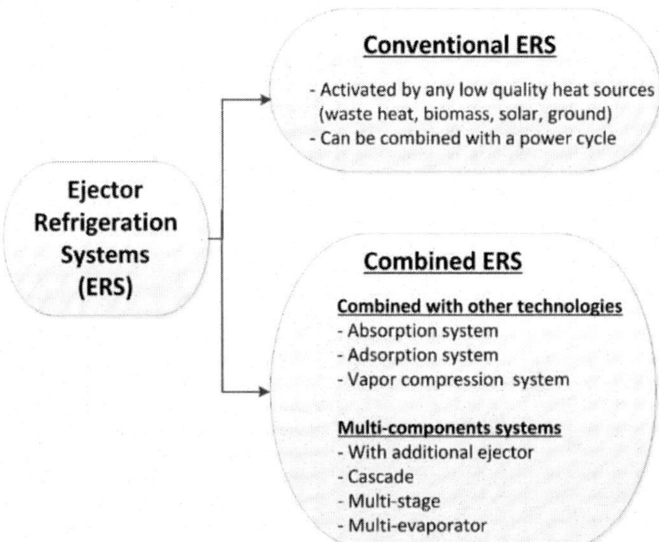


Figure 10. ERS classification.

Cycles and systems for heating and refrigeration as shown in Figure 10 are extensively discussed in the literature. Interested readers are referred to (Jiwei Yan et al. 2017; García Del Valle et al. 2014; S. He, Li, and Wang

2009; Selvaraju and Mani 2006; Besagni, Mereu, and Inzoli 2016; Aidoun et al. 2019a) for further details.

The developments that will follow will rather be devoted to the application potential and versatility of ejectors, in stand-alone or combined with other technologies for the recovery of waste heat.

## 6.2. Ejector-Absorption Systems

Combined ejector-absorption cycles or other ejector boosted absorption set-ups may substantially improve overall efficiency. Such systems have received a great deal of attention for their positive impact on performance in exchange of relatively minor modifications to their original configurations. Many investigations continue to be conducted on this well documented subject, mostly concentrated in refrigeration and heating applications. Typically recent information can be found in the following references (Sirwan et al. 2013; Abed et al. 2015; Shi et al. 2018; Kumar and Kumar 2014; I.W Eames, Aphornratana, and Haider 1995; D.-W. Sun, Eames, and Aphornratana 1996; Satha Aphornratana and Eames 1998; Wu and Eames 1998; Ian W. Eames and Wu 2000; Alexis and Rogdakis 2002).

More generally, for systems involving complex combinations of ejector absorption systems, solely activated by a heat source, the work of S. Zhang et al. (2018) for dry freezing may be given as an example. In this system, represented in Figure 11, high-pressure steam, generated from waste heat recovery, successively drives a multi-stage ejector vacuum pump, an ammonia-water absorption refrigerator for freezing foods and cooling hygroscopic solution cycled to absorb sublimated water vapor from frozen stuff. According to a thermodynamic analysis performed by the authors, the proposed system heat requirements are up to 46% lower compared to a conventional system driven by electricity produced by the steam at the same pressure.

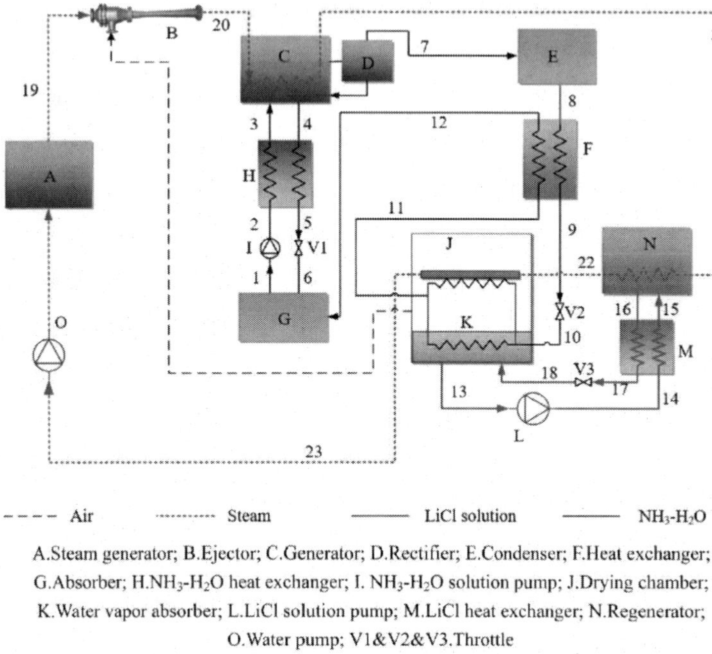


Figure 11. Schematic diagram of a new freeze-dryer (S. Zhang et al. 2018).

## 6.3. Ejector-Compressor Heat Pumps

There are many ejector-compressor combinations for heat pump cycles available in the literature and extensively discussed in many recent reviews. For example, the cascade cycle of Figure 12 proposed by Megdouli et al. (2016) consists in introducing an ERS loop using R134a as refrigerant in a mechanical compression $NH_3/CO_2$ transcritical cascade. The ERS loop is heat activated by the gas cooler. A theoretical performance comparison by the authors suggests that for a given cooling capacity, COP and second law efficiency can increase by up to 37% and 12%, respectively, over those of the conventional cascade cycle under the given operating conditions at the optimum gas cooler pressure.

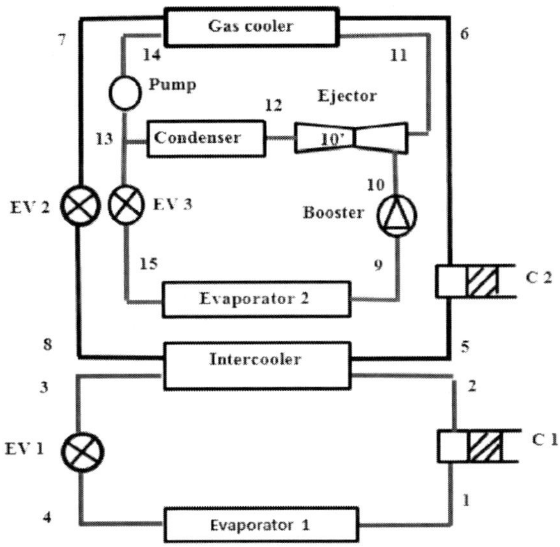


Figure 12. Schematic of the hybrid cascade refrigeration cycle (Megdouli et al. 2016).

Later, F. Li et al. (2018) proposed a cascade cycle consisting of an ejector loop driven by solar energy with R134a, R1234yf or R141b as working fluid and a mechanical compression loop with refrigerant R1234yf (Figure 13). The temperature levels investigated were -20°C in the evaporator and 45°C in the condenser. Energy and exergy analyses applied in these conditions resulted in the following conclusions:

- The exergy destruction occurs principally in the solar collector-generator, the ejector, the compressor and the expansion valve.
- COP of the solar cascade system is 45.8% higher than for a conventional air source heat pump.
- The capacity of the solar cascade system is 95.3% higher than for a conventional air source heat pump.

It must however be pointed out that the conventional cycle serving for comparison is single stage and the solar gain in the ejector cascade is considered as free energy.

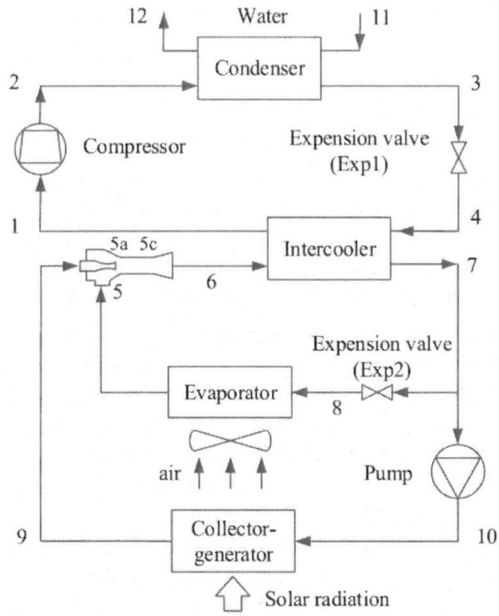


Figure 13. Solar-driven ejector-cascade heat pump cycle (F. Li et al. 2018).

Refrigerant mixture properties were put to use in many cases to take advantage of their temperature glide for heating. In this context a hybrid refrigeration system consisting of auto-cascade refrigeration cycle (HACRS), coupled with a heat-driven ejector cooling cycle was employed by Hao et al. (2018) and shown in Figure 14. A refrigerant mixture of R170/R600 was theoretically investigated, to successively benefit both loops of the system from the thermophysical properties and the temperature glide of the mixture. The ejector loop was thermally activated with heat recovery. The authors' findings indicated that the compressor energy consumption in this set up could be reduced by 50% with respect to the conventional auto-cascade refrigeration cycle.

Another solar activated cycle was proposed by Xu et al. (2018) and represented in Figure 15. Its main difference with the conventional mechanical-thermal ejector cycle commonly described in the literature is the addition of a second expansion valve and a separator to the system. The separator located immediately after the first expansion valve receives two-phase condensate at an intermediate pressure. Vapor is drawn by the ejector and the liquid is further expanded to the evaporator conditions prior to being handled by the compressor. Both compressor and ejector outlet streams merge at the same pressure state before returning to the condenser to close the cycle. Performance estimates of the new cycle with R152a suggest 22.8% preliminary improvements in terms of COP over the conventional hybrid cycle.

Lately (Fan, Yan, and Yu 2019) modified a solar assisted heat pump cycle to use a zeotropic mixture of refrigerants R290/R600a for water heating applications (Figure 16). The heat pump was designed to use air and solar energy as heat sources. The use of zeotropic mixture characterized by a temperature glide during the heating process enables the refrigerant to maintain good flow temperature matching in the condenser with water temperature variation and consequently improving heat transfer efficiency. A parametric study conducted by the authors to assess the new system behaviour showed that for condenser inlet temperatures in the 45-80°C range, COP and heating capacity may be up to 33% and 47% respectively over the conventional heat pump at a fixed evaporator outlet temperature of 10°C.

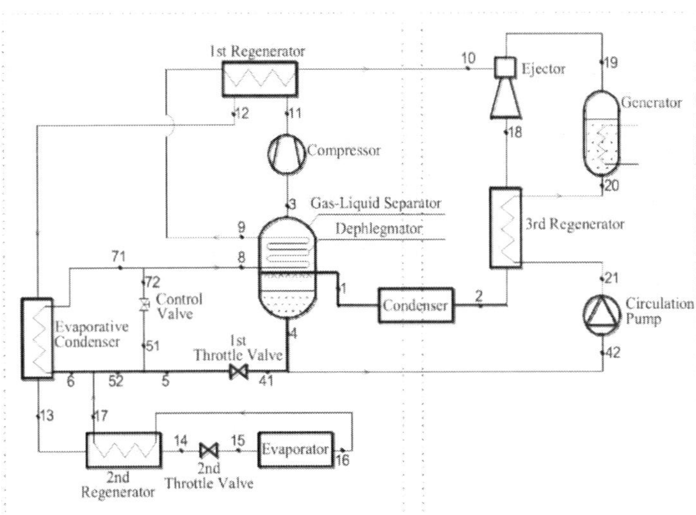


Figure 14. Schematic diagram of the HACRS (Hao et al. 2018).

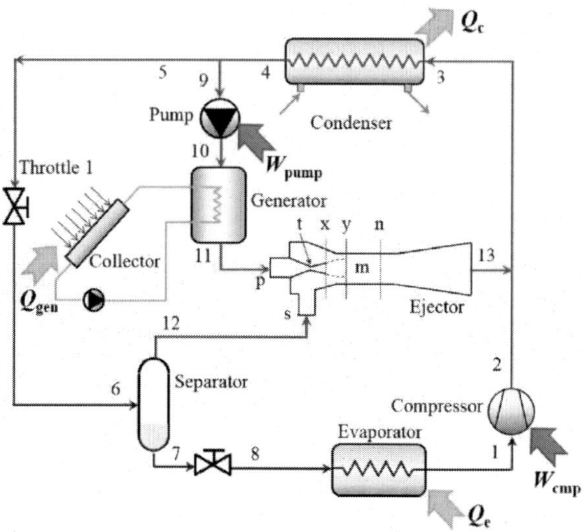


Figure 15. Solar activated ejector-compressor hybrid system (Xu et al. 2018).

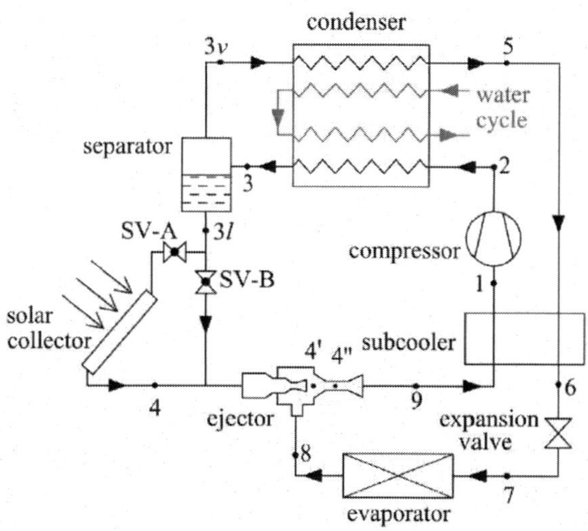


Figure 16. Schematic diagram of the mixture heat pump (MHP) (Fan, Yan, and Yu 2019).

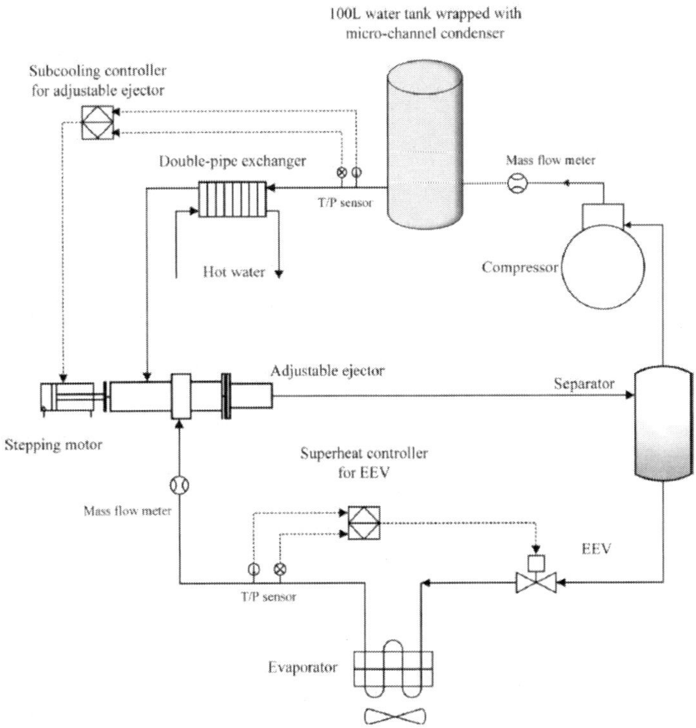


Figure 17. Schematic diagram of the DX-SEHPC experimental setup (Jiaheng Chen, Yu, and Qian 2019).

A recent theoretical and experimental work worth mentioning is on solar assisted ejector compression heat pumps for water heating, published by Jiaheng Chen, Yu, and Qian (2019). The authors' assessment particularly focusing on the development of a control method capable of tracking the degree of sub-cooling and adjusting accordingly the nozzle throat area of the ejector to vary the flow rate. Tests performed on the ejector-expander heat pump represented in Figure 17 indicated that the sub-cooling degree could be maintained in a range of 4-7°C, reducing the power consumption by up to 6.99%, in comparison with those with fixed throat geometry.

## 6.4. Ejector-Transcritical Cycle Integration

Several theoretical and experimental studies have been devoted to using the ejector as an expander in transcritical $CO_2$ refrigeration and heat pump systems (Boccardi et al. 2017; Zheng and Deng 2017; Y. He et al. 2017).

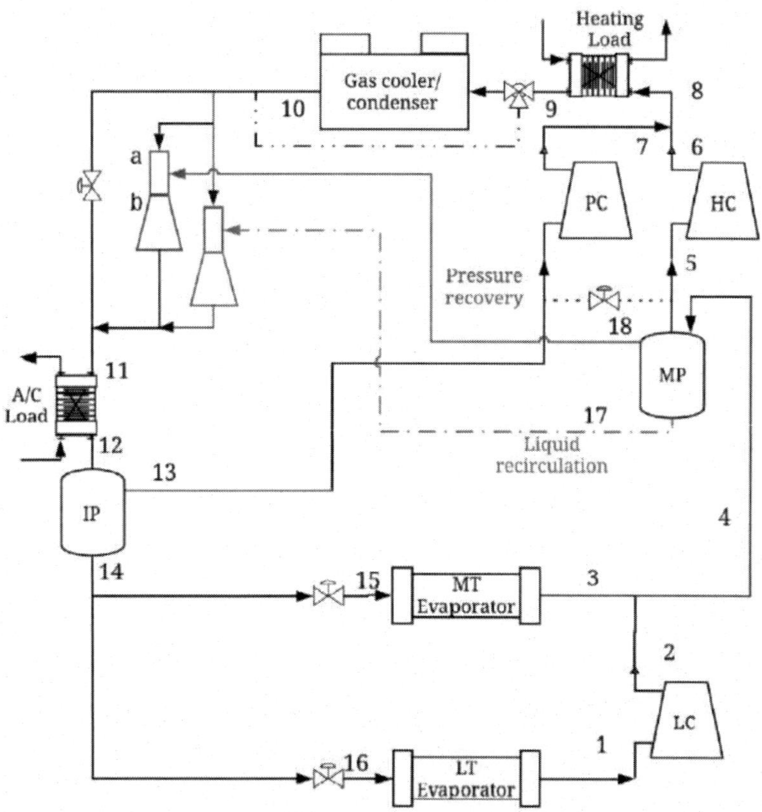


Figure 18. Transcritical $CO_2$ booster refrigeration system with multi-jet ejectors (Purohit et al. 2018).

An integrated $CO_2$ booster refrigeration unit with multi-jet ejectors for supermarkets, represented in Figure 18 was proposed by Purohit et al. (2018). Multi-ejectors in this system were a combination of vapor and liquid devices of different sizes, connected in parallel. Vapor ejectors ensured the stability of the gas cooler pressure, entrained part of the refrigerant from the medium pressure receiver (MP) and discharged in the higher-pressure receiver (IP), thereby relieving the compressor (HC) from partial capacity and reducing its power consumption. Liquid ejectors transferred part of the liquid in the receiver (MP) to higher-pressure receiver (IP), thereby correctly overfeeding evaporators (MT). The compressor (PC) removed flash gas in the receiver (IP). The authors advised that this configuration was suitable for cold and mild climates where its performances in terms of COP and total $CO_2$ equivalent emissions were respectively 20% higher and 63% lower than those of the baseline R404A booster were.

## 6.5. Ejector District Heating Enhancement

In order to encourage more waste heat use in district heating networks, its economical transmission over long distances must be improved. This may be achieved by lowering the return temperature of the primary heating network (PHN). In this respect, recent research of F. Sun et al. (2018) based on the concept of ejector heat exchanger system, proposed two configurations based of the ejector refrigerator with a pressure booster, respectively depicted in Figure 19. In Figure 19a, the pressure booster located on the refrigerant line, between the ejector outlet and the condenser inlet. In the second configuration, the pressure booster located in the refrigerant line between ejector secondary fluid inlet and the evaporator refrigerant outlet, as shown in Figure 19b. The thermodynamic analysis of both configurations indicated that the scenario of Figure 19a decreased the booster pressure and its power consumption while improving the cycle exergy efficiency. Lower return water temperatures could be obtained, substantially increasing the economical heat transmission distance of the primary heating network.

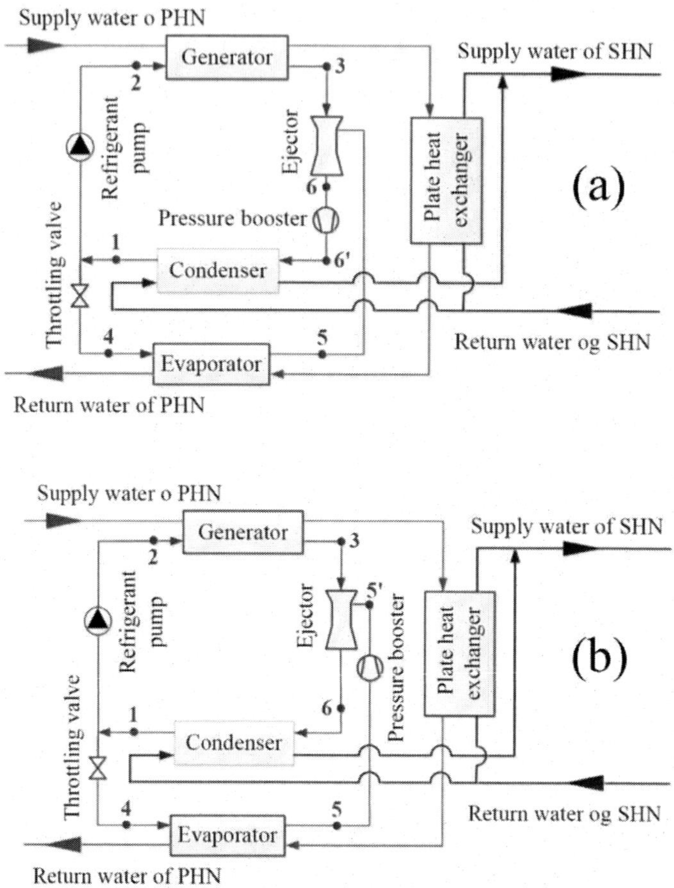


Figure 19. The two configurations of ejector-district heating systems in (F. Sun et al. 2018).

# 7. Ejectors in Power Systems

## 7.1. Ejector-ORC Cycles

Several works on ejector enhanced ORCs activated with waste heat recovery or renewable sources are available in the literature.

In the configuration proposed by X. Li, Li, and Zhang (2013) (Figure 20), the primary flow of the ejector is provided by part of the liquid successively leaving the condenser, pumped to the primary pressure through a heat exchanger boiler using the degraded heat stream leaving the ORC boiler. The vapor generated in this way activates the ejector to draw the flow expanded in the turbine as a secondary stream. This arrangement has the potential of enhancing flow expansion in the turbine, thus increasing its power output.

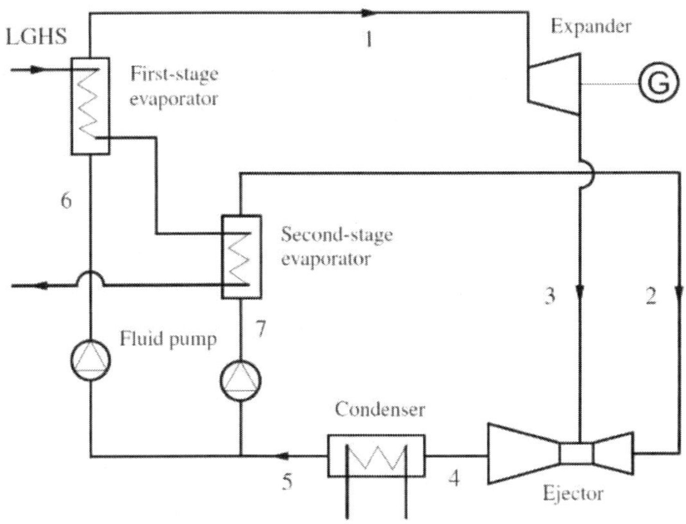


Figure 20. Ejector-ORC system in (X. Li, Li, and Zhang 2013).

In more recent work, Haghparast et al. (2019) analyzed the effects of ejector operating parameters on the ORC performance, with R245fa as working fluid. Results showed that the diameter of the constant area-mixing chamber is one of the most important geometrical parameters, influencing the power generation and the cycle performance. The power output increases with secondary mass flow rate, and decreases with increasing throat diameter, primary inlet pressure, or primary inlet temperature.

Mondal, Alam, and De (2018) proposed an organic flashing cycle (OFC) integrating an ejector as presented on Figure 21. The original cycle of the basic OFC is characterized by high irreversibility in the throttling devices, due to high pressures required in this process, such that the gains of better temperature matching become neutralized. An ejector is thus, used instead at the turbine outlet in replacement of the low-pressure valve. In this configuration, high-energy liquid from the separator activates the ejector, which draws the vapor stream leaving the turbine. The ejector favors a lower turbine outlet pressure resulting in a higher power output. Cycle efficiency and power output are expected to improve by up to 9.5% according to this evaluation.

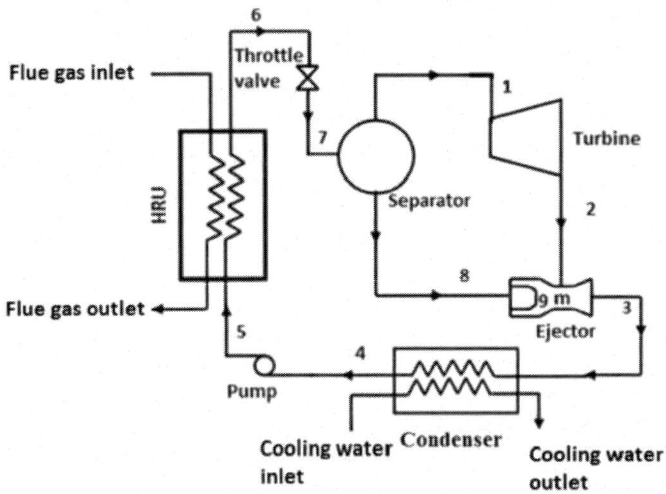


Figure 21. Layout of an OFC with ejector (Mondal, Alam, and De 2018).

Further, many researchers have studied $CO_2$ transcritical ORC and the main difficulty remains the $CO_2$ condensation at relatively high temperature, due to the low critical temperature of this fluid. Figure 22 employing an ejector is proposed by Xia et al. (2018) in order to induce transcritical $CO_2$ ORC with no additional complexity to the system. A parametric analysis conducted by the authors demonstrated the feasibility and the conditions for optimal operation of this system.

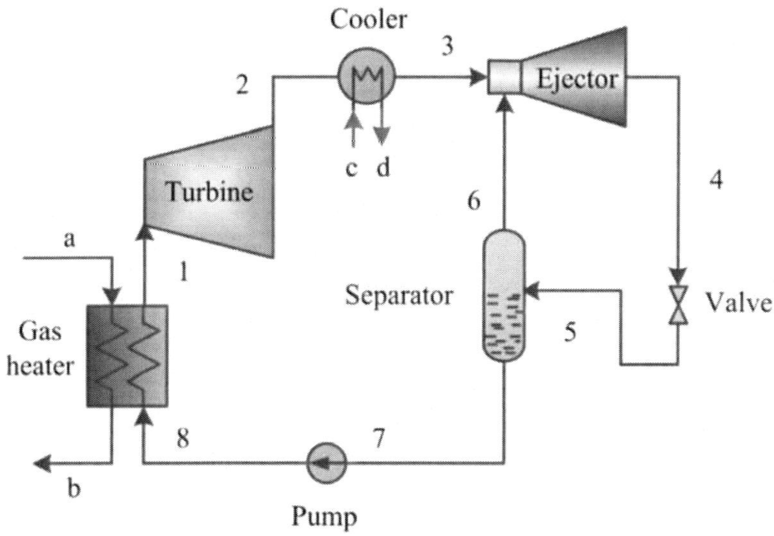


Figure 22. Schematic diagram of the novel transcritical $CO_2$ Rankine cycle with an ejector (Xia et al. 2018).

## 7.2. Ejector-Kalina Cycles

Integrated ejector-Kalina cycles for power generation and refrigeration effect are so far little studied. However, some researchers have begun showing interest in exploring this avenue by combining ERS and Kalina cycles in order to form new efficient systems, as the many found in the literature with ORC, Rankine and other heat pump configurations.

For example, among these works, X. Li, Zhang, and Li (2013) replaced the throttling valve and the absorber in the conventional Kalina cycle by a two-phase ejector, as seen in Figure 23 above. In this way, the power output and the thermal efficiency of the cycle both increase, due to the ejector increasing the expansion ratio of the turbine, in comparison to the conventional cycle, without affecting the condenser.

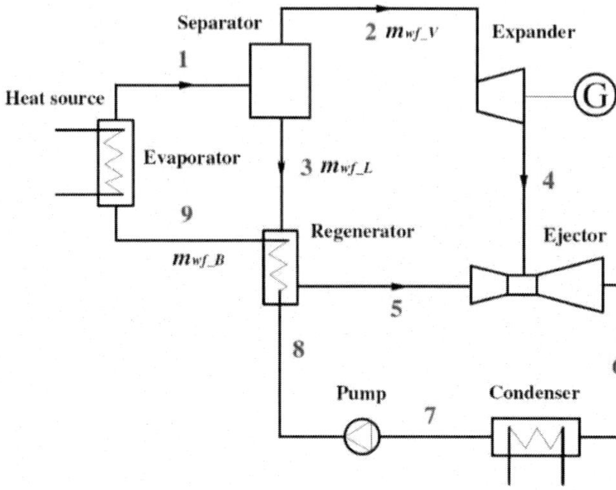


Figure 23. Kalina cycle with ejector (X. Li, Zhang, and Li 2013).

## 7.3. Ejector-Brayton Cycles

Supercritical $CO_2$ (S-$CO_2$) Brayton cycles of high thermal efficiency with elevated temperature heat sources has been identified as an alternative to the conventional Rankine cycle, typically less compact and comparatively less efficient. Thus, this new cycle is proposed as a suitable application for nuclear and concentrated solar thermal (CST) processes. Padilla et al. (2016) proposed three potential S-$CO_2$ Brayton cycle configurations integrating an ejector for the CST case. Figure 24 represents the first configuration along with a reference Brayton cycle. The role of the ejector is to increase the pressure before the heater and allow the solar receiver to operate at lower S-$CO_2$ pressures without additional energy consumption and eventually improving the CST thermal efficiency, cost and durability. The authors theoretically analyzed the three configurations using thermodynamic and multi-objective optimization procedures, which they compared with the corresponding conventional S-$CO_2$ Brayton cycles. The proposed configurations are shown to achieve at some conditions higher thermal efficiency than the corresponding steam Rankine cycles and approach the conventional S-$CO_2$ Brayton cycles. Given the high ejector design characteristics in terms of entrainment and compression ratios necessary to operate in the context of S-$CO_2$ cycles considered by the authors, Palacz et al. (2018) used the same conditions to redesign the ejector by means of a CFD tool for the first configuration. The results of this investigation revealed that a pressure lift of 103 bar and a mass entrainment ratio of 0.995 were necessary in order to ensure the thermal efficiency of the system, showing that this requirement cannot be achieved in current ejector practice.

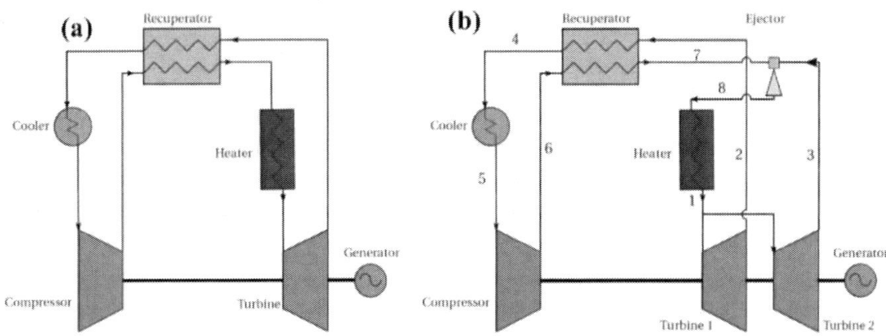


Figure 24. Comparison of the Brayton cycle (a) without ejector and (b) with ejector (Palacz et al. 2018).

## 7.4. Ejector-Fuel Cells Systems

Ejector technology may be used to replace the pumping recovery devices of the hydrogen not consumed by fuel cells systems. Such replacement saves the electricity consumed by the pumping device and contributes to improve the overall efficiency of the plant (Zhu, Li, and Cai 2011; Hwang 2014; Rao and Jagadeesh 2015; Maghsoodi, Afshari, and Ahmadikia 2014).

More particularly, it can play an important controlling role in the operation and efficiency of hybrid solid oxide fuel cell-gas turbine systems through efficient, safe and reliable anode and cathode fluid recirculation. At part load, temperature control by means of ejectors monitoring the difference between anode and cathode channels with variable-speed operation was shown by (Jinwei Chen, Li, et al. 2018; Jinwei Chen, Chen, et al. 2018) to significantly influence system efficiency and stability. The process layout is shown in Figure 25. Anode and cathode streams circulate simultaneously in two distinct loops. In the anode stream, preheated natural gas mixed with recirculated anode exhaust by a first ejector becomes hydrogen rich before further heating by gases of an afterburner. In entering the anode channel, an electrochemical reaction takes place. In the cathode stream, compressed and preheated air mixes with the cathode exhaust by means of a second circulation ejector before entering the cathode channel. Excess fuel from the anode-cathode compound goes to the after-burner to extract the remaining energy. These partly heat inlet mixed fuel to the anode and activate the turbine for electricity generation. The turbine gas outlet preheats inlet air and natural gas to the process.

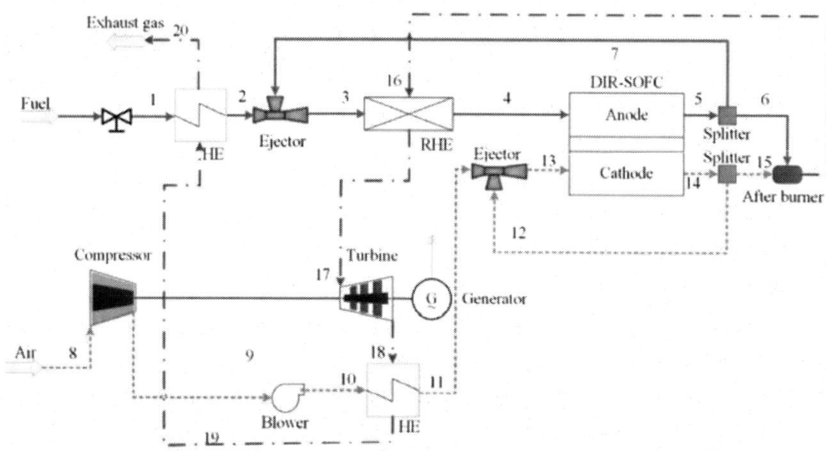


Figure 25. The integration scheme of the SOFC-GT system (Jinwei Chen, Chen, et al. 2018).

Recirculation in high temperature fuel cells is a critical performance and safety issue. Therefore, a systematic theoretical analysis based on CFD techniques was conducted by Ferrari, Pascenti, and Massardo (2018) on the application of such ejectors for integration in high temperature fuel cells, considering that accurate ejector design for these applications is of crucial importance. Further work on CFD-based ejector design for integration in proton membrane fuel cell systems is available in the recent literature in the context of internal combustion engine replacement (Besagni et al. 2017), as well as for the case of forklift truck ejector-fuel cell design (Hosseinzadeh et al. 2014).

# 8. EJECTORS IN TRI-GENERATION AND OTHER MISCELLANEOUS SYSTEMS

## 8.1. Ejectors in ORC-Refrigeration Systems

Combined ejector and Rankine system (Oliveira et al. 2002) represented in Figure 26 was designed with the ejector and the turbine set in parallel. They work with the same working fluid conditions fixed by the generator. In this way, the turbine generates power and the ejector a refrigeration effect, separately. Tests with n-pentane have shown that at evaporation temperature of 15°C, condensation temperature of 35°C and when the working fluid inlet temperature was at 95°C, the refrigeration loop COP reached 0.3 and the power generation loop had an efficiency of 3-4%. In order to maintain operation stability in this system, however, fine-tuned control of outlet conditions of both the ejector and turbine need to be established for a good pressure balance.

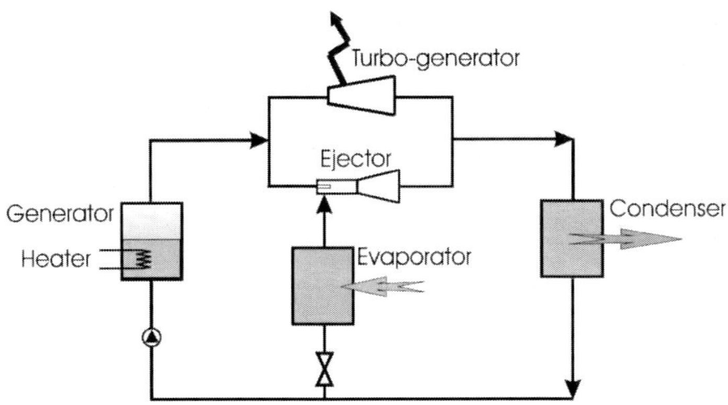


Figure 26. Ejector-ORC system (Oliveira et al. 2002).

A simple combination of an ORC and an ejector loops to form a hybrid system is represented in Figure 27 (Dai, Wang, and Gao 2009). Turbine outlet flow activates the ejector as the primary stream, which draws the secondary stream from the evaporator of the ejector loop. Vapor stream at the ejector outlet leaves the condenser as liquid and split between ORC and ejector loops.

Habibzadeh, Rashidi, and Galanis (2013) conducted a parametric study on a combined power and ejector-refrigeration cycle driven by waste heat with several refrigerants. In their configuration, the ejector is activated by bleeding some partially expanded working fluid from the turbine. The fully expanded portion of the remaining fluid in the turbine mixes irreversibly with the ejector outlet stream, prior to the pre-heater (Figure 28).

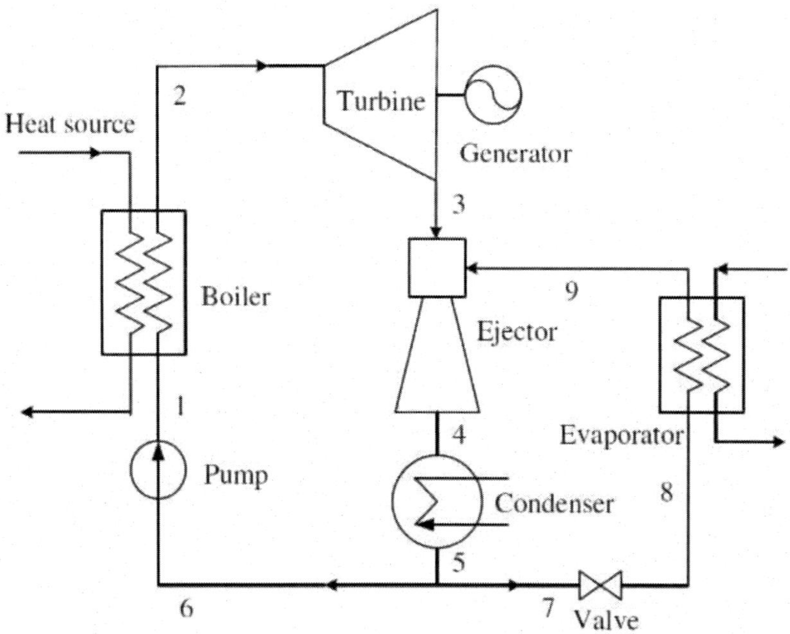


Figure 27. Ejector-ORC system (Dai, Wang, and Gao 2009).

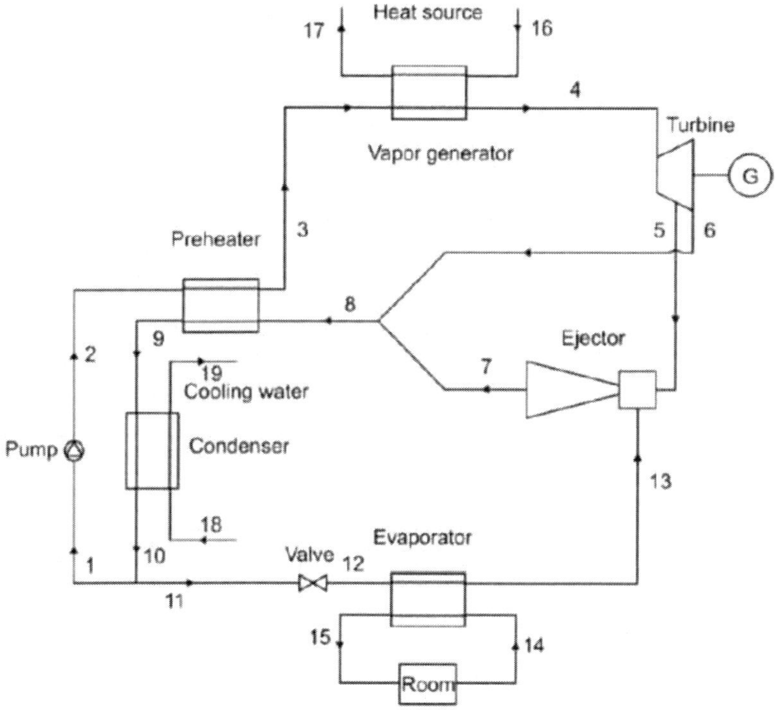


Figure 28. Schematic diagram of the combined power and refrigeration cycle (Habibzadeh, Rashidi, and Galanis 2013).

## 8.2. Ejectors in Kalina-Refrigeration Systems

More recently, Ghaebi et al. (2018) proposed a new cycle, driven by a geothermal heat source. The Figure 29 representing this cycle combines an ERS loop and a Kalina cycle loop for power generation, integrated into a unique system. According to the energy and exergy analyses conducted by the authors on the effect of key operating parameters and conditions on performance, it was possible to optimize the conditions for a superior turbine power output in comparison to that of the conventional system, in addition to providing a higher refrigeration effect.

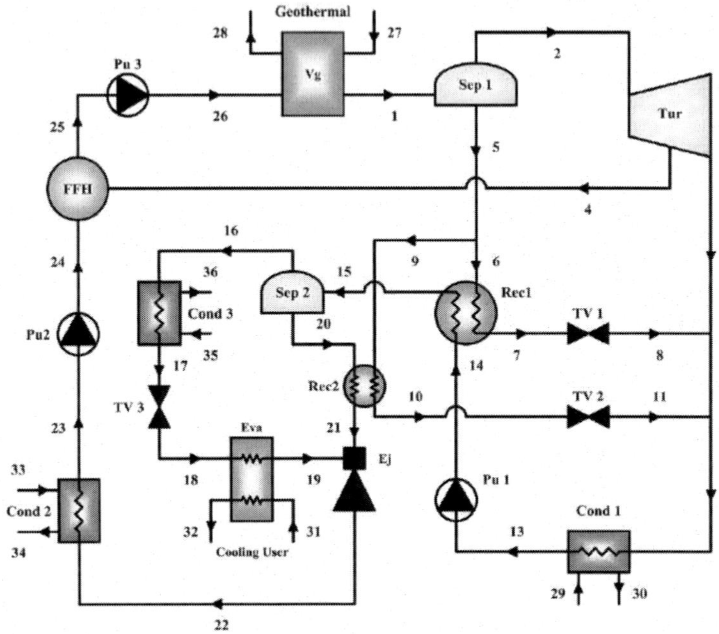


Figure 29. Schematic diagram of the combined ERS and the Kalina cycle (Ghaebi et al. 2018).

## 8.3. Ejectors in CCHP Systems

Combined cooling, heating and power systems (CCHP) have the flexibility to use different technologies and working fluids to produce simultaneous heating, cooling and power. They can integrate devices and systems such as ejectors, absorption and micro-steam turbine loops powered by natural gas as studied by Ebrahimi, Keshavarz, and Jamali (2012) (Figure 30) or use organic fluids and operate on low-grade energy as will be discussed in the present paragraph.

Boyaghchi and Heidarnejad (2015) assessed a combined cooling, heating and power (CCHP) system, solar activated and integrating an ERS as well as an organic Rankine cycle loops. The system represented in Figure 31, was analyzed in terms of thermodynamics and thermoeconomics considerations for an optimized operation for winter and summer building applications. The results indicated that the summer thermal and exergy efficiencies were 28% and 27%, respectively, while in the winter, they were 4% and 13%.

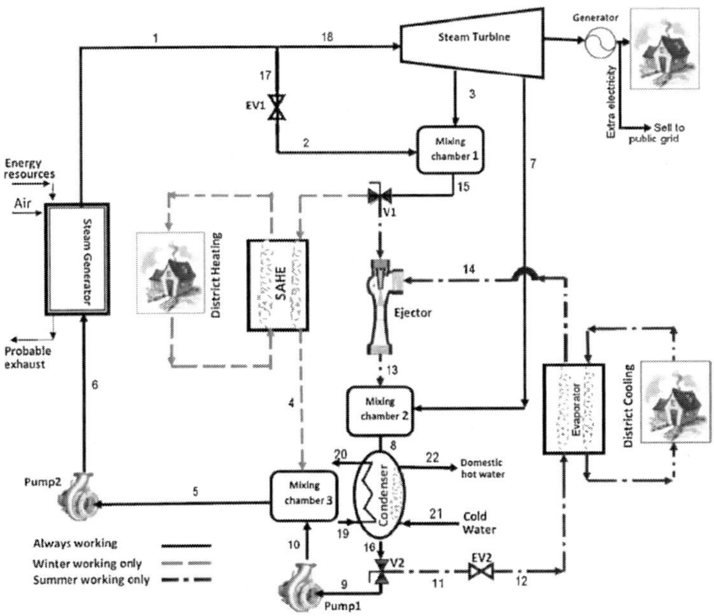


Figure 30. Schematic diagram of the CCHP cycle (Ebrahimi, Keshavarz, and Jamali 2012).

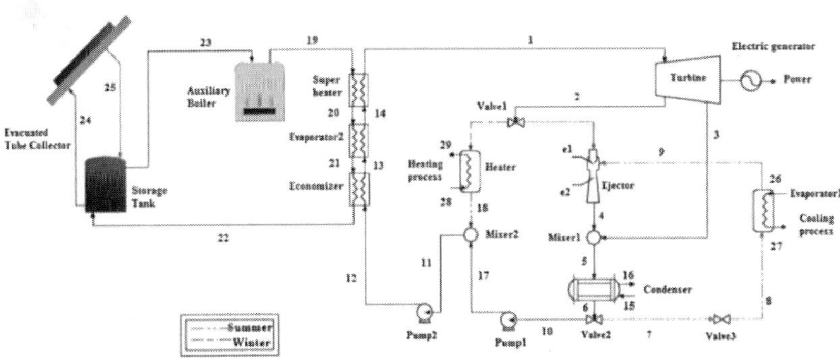


Figure 31. Solar micro-CCHP layout (Boyaghchi and Heidarnejad 2015).

Yang et al. (2016) combined power and ejector refrigeration cycle used a zeotropic refrigerant mixture (isobutane/pentane) having different compositions in the refrigeration and power loops connected by a vapor-liquid separator. A comparison with a conventional combined power and ejector-refrigeration cycle revealed that the new cycle generated more power but less refrigeration effect with a better exergy efficiency. In addition, the system performance was higher at lower condensing temperature. Refrigeration performance improved at higher generation temperatures but the power cycle performance deteriorated (Figure 32).

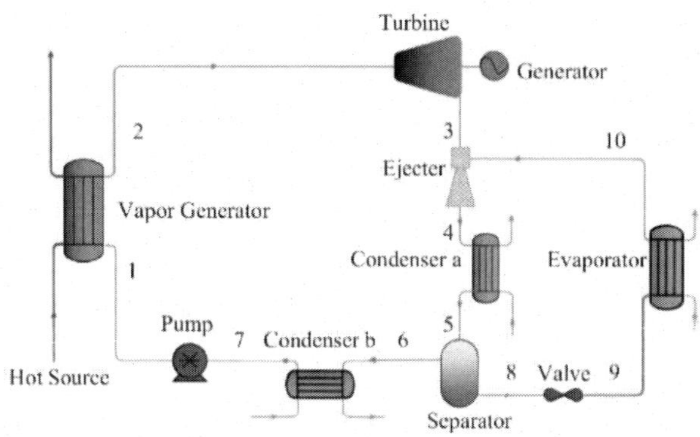


Figure 32. Combined power and ejector-refrigeration cycle (Yang et al. 2016).

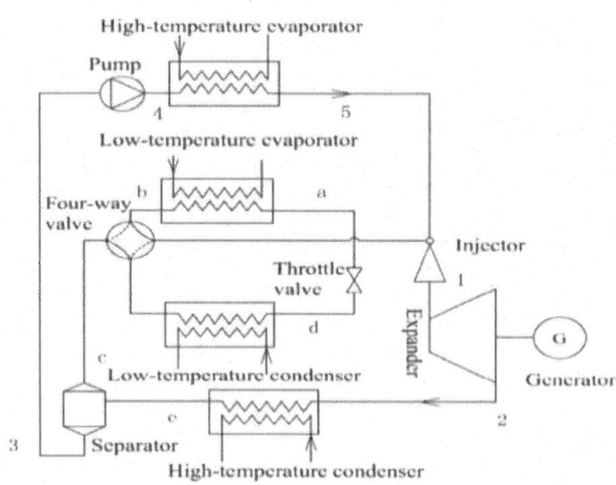


Figure 33. Structure of CCHP–ORC system installed with heat pumps using zeotropic mixtures (Z. Li, Li, and Xu 2016).

Along similar lines, further research by (Z. Li, Li, and Xu 2016) was performed on a combined cooling, heating and power ORC system (CCHP-P) depicted in Figure 33.

It consisted of an ORC subsystem using zeotropic mixtures and a heat pump subsystem. The system might as well, be used for combined cooling and power (CCP) or combined heating and power (CHP) by adjusting a four-way valve. In addition, this design could handle geothermal water at 95°C with a mass flowrate of 40 t/h. The heat transfer fluid was heated to 45°C for heating with the ambient temperature at -5°C, and the refrigerating fluid was cooled to 0°C with the ambient temperature at 35°C. Several refrigerant mixtures were evaluated among which R141b/R134a, R141b/R152a and R123/152a pairs showed higher COP and exergy efficiency. A closer analysis of these zeotropic pairs in terms of concentrations, led the authors to infer that a high proportion of wet working fluid and a low proportion of dry working fluid in the mixture conditioned higher system performance.

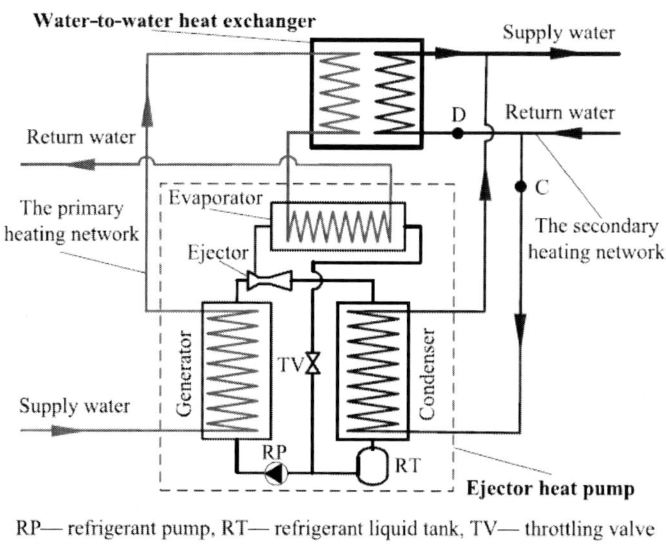


Figure 34. District heating with ejector heat pump (F. Sun et al. 2014).

Other authors (K. Zhang et al. 2016) rather focused their efforts on evaluating losses in ejectors designed for ORC systems and how they reacted to parameters changes and interactions within the system.

The use of ejectors to enhance heat transfer efficiency in district heating networks is illustrated by the work of F. Sun et al. (2014). An ejector-CHP system shown in Figure 34 was first analyzed. In this particular case, waste heat district heating system with CHP, based on ejector heat exchangers and absorption heat pumps was compared with a conventional district heating system with CHP. The ejector set-up was activated by a steam turbine exhaust; and according to the authors' analysis the steam consumption could decrease by 41.4% and heat transmission capacity increase by 66.7% without changing the flow rate of circulating water in the network.

## 8.4. Ejectors in Absorption-Power Systems

An interesting application of an ejector enhanced absorption power cycle (APC) coupled to an electrolyzer for water separation in hydrogen and oxygen was proposed by Yosaf and Ozcan (2018).

In Figure 35, a binary mixture of LiBr-Water is preheated and pumped to the generator pressure where it is further heated by waste heat. The flow is then split in the separator with the steam directed to the turbine for power production and a strong solution to activate an ejector, which in turn, draws the expanded steam leaving the turbine. The ejector delivers a mixed and pressurized stream (steam + strong solution) to the absorber to close the loop. The proton-exchange membrane electrolyzer (PEME) uses pumped water and the generated electricity to produce $H_2$ and $O_2$. The ejector extends the expansion in the turbine, enhancing power production without decreasing the absorber temperature, thus enhancing overall system efficiency and higher hydrogen production.

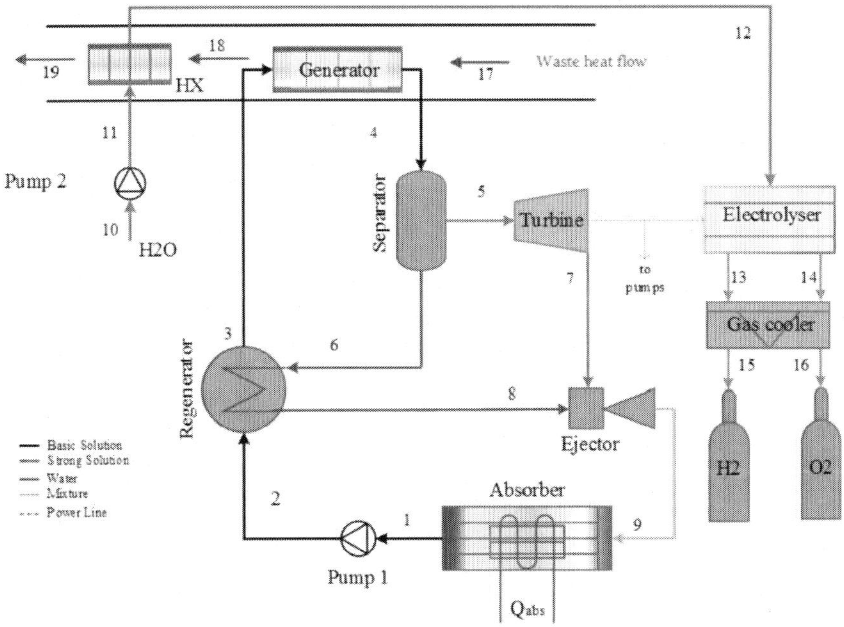


Figure 35. Ejector absorption-power cycle (Yosaf and Ozcan 2018).

A further potential application is the hybrid solar activated system investigated by Khaliq, Kumar, and Mokheimer (2018), which includes an ORC working with R141b, R600a, R290, R717 and R143a and an ejector-absorption refrigeration loop using $NH_3$-$LiNO_3$ as the working fluid for electricity and refrigeration (Figure 36). The solar loop is designed to generate a high temperature level for power generation in ORC and a low temperature level for refrigeration. The parametric study conducted by the authors in terms of energy and exergy analyses concluded that the solar beam, the inlet pressure to the turbine, temperatures at the parabolic trough collector (PTC) and the ORC pump as well as the working fluid considerably affect energetic and exergetic outputs of the overall system.

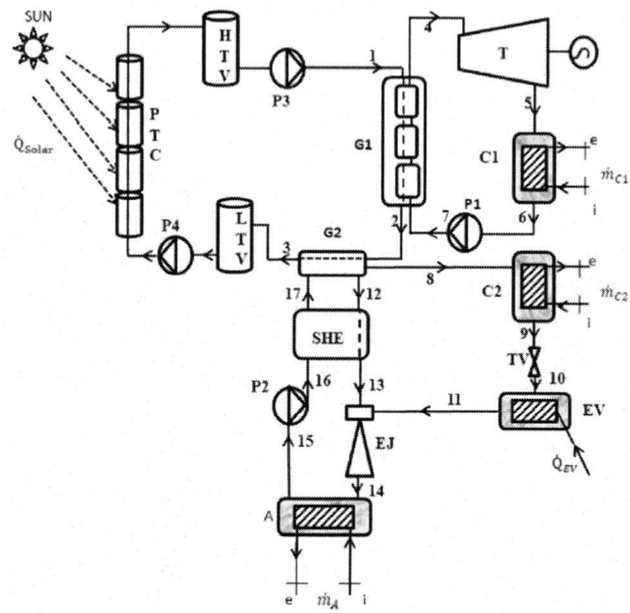


Figure 36. Schematic diagram of a solar assisted combined power and refrigeration system (Khaliq, Kumar, and Mokheimer 2018).

## 8.5. Combined Ejector-ORC in Desalination Plants

Among the current desalination technologies, the multi-effect thermal vapor compression approach (MED-TVC) attracts more attention because of the negative pressure distillation environment in which seawater boiling takes place, thus facilitating its evaporation. Under these conditions, most types of low-grade heat may be used and steam ejectors are particularly suitable for MED-TVC applications since they favor non-condensed steam circulation between the desalination effects while maintaining the whole system under the required vacuum. A MED-TVC system schematically is depicted in Figure 37 and Tang et al. (2018b; 2018a) described its operation details.

C. Li et al. (2012) proposed another configuration worth mentioning and presented in Figure 38. This system combines MED desalination with a supercritical organic Rankine cycle (SORC), an ejector and operates like a combined heat, power and condensation system. The ejector condenses the vapor of MED's final effect and the SORC provides the necessary power to the desalination plant. The ejector-ORC loop uses R152a as working fluid so that a low-grade heat source may be used to activate it. The authors propose the use of a pressure exchange ejector, which up to now has not been tested experimentally.

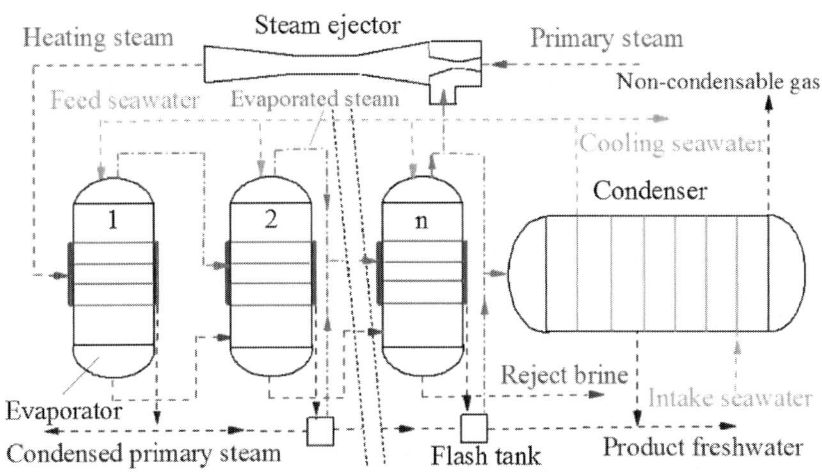


Figure 37. Schematic diagram of MED-TVC desalination system (Tang et al. 2018b).

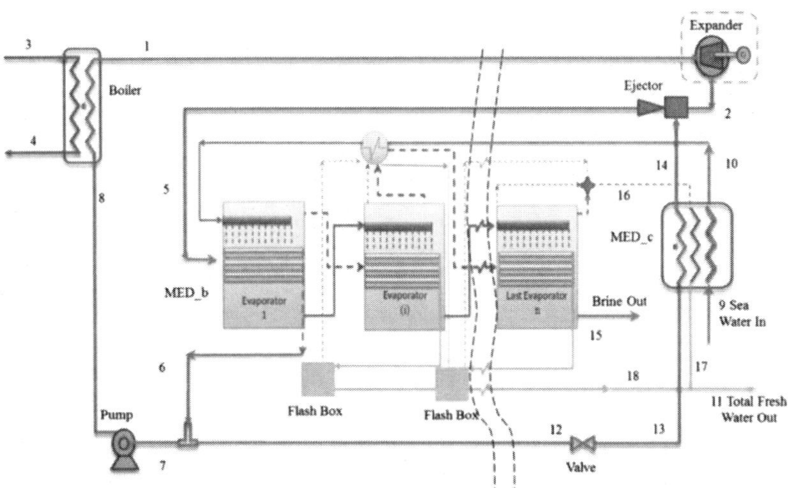


Figure 38. Schematic of the proposed SORC-Ejector-MED system (C. Li et al. 2012).

## CONCLUSION

This chapter introduces the ejector technology, which draws particular attention, due to its many advantages including the ability to recover waste heat at the lower range of temperatures, its simplicity and high flexibility for use alone or in combination with many available technologies in industrial and building applications. Careful attention is paid to the various ways it can contribute to waste heat recovery through combinations with other technologies, integrated in systems or used as stand alone equipment. More particularly, ejectors have the potential to enhance a variety of waste heat recovery technologies, especially with low waste heat grades in refrigeration-heating, power and combined systems.

The main concluding elements emerging from this assessment may be stated as follows:

- Ejector is a simple device, well adapted to recover waste heat at low temperature for its activation. This feature coupled with simple geometry is a very interesting asset opening a range of application possibilities for this technology. In view of the numerous studies of ejectors in combination with other technologies, integrated in processes and systems, the WHR potential of ejectors deserves more consideration.
- The technology of ejectors still remains relatively unknown to potential users and little exploited. Yet, it offers great opportunities for thermal energy upgrading, refrigeration and heating. Promising applications are expected for waste heat upgrading in industry by means of hybrid systems or by combined use with thermal solar energy in buildings.
- Rather than considering this technology to compete with mechanical compression (due to its comparatively low cycle efficiency), it must be seriously examined in combination (not competition) with most WHR technologies in hybrid systems for heat upgrading, power generation and other benefits.

- Ejector industrial applications are limited to steam in niche applications (heat upgrading, vacuum creation, process mixing…). Experimental studies of the ejectors are mostly related to refrigeration and heating needs. In industrial applications with working fluids other than steam, they are still scarce but, due to their versatility and relative ease of combined operation with other WHR technologies one can foresee increased progress in the future. Many theoretical investigations of high quality are already available at the industrial size scale and discussed.
- Generally, low grade recovered waste heat is the source of single-phase ejector activation, which may be used to provide compression, cooling or heat upgrading in combined systems. Two-phase ejectors are integrated as auxiliary components in cycles and processes to enhance performance (by reducing throttling losses, regulating fluid circulation…).

## Acknowledgments

The authors would like to acknowledge the support of CanmetENERGY-Varennes (Natural Resources Canada), and Energy Technologies Laboratory (Hydro-Québec).

## References

Abdulateef, J. M., K. Sopian, M. A. Alghoul, and M. Y. Sulaiman. 2009. "Review on Solar-Driven Ejector Refrigeration Technologies." *Renewable and Sustainable Energy Reviews* 13 (6–7): 1338–49. https://doi.org/10.1016/j.rser.2008.08.012.

Abed, Azher M., M. A. Alghoul, R. Sirawn, Ali Najah Al-Shamani, and K. Sopian. 2015. "Performance Enhancement of Ejector-Absorption Cooling Cycle by Re-Arrangement of Solution Streamlines and Adding

RHE." *Applied Thermal Engineering* 77: 65–75. https://doi.org/10.1016/j.applthermaleng.2014.12.003.

Abed, Azher M., K. Sopian, M. A. Alghoul, Hasan Sh., and Ali Najah Al-shamani. 2017. "Experimental Evaluation of Single Stage Ejector-Absorption Cooling Cycle under Different Design Configurations." *Solar Energy* 155: 130–41. https://doi.org/10.1016/j.solener.2017.06.005.

Ablwaifa, Ali, Ian W Eames, and V O Petrenko. 2009. "Experimental Validation of CFD Model Used to Design Jet-Pumps." In *International Seminar on Ejector/Jet- Pump Technology and Application*. Louvain, Belgium.

Aidoun, Z., K. Ameur, M. Falsafioon, and M. Badache. 2019a. "Current Advances in Ejector Modeling, Experimentation and Applications for Refrigeration and Heat Pumps. Part 1: Single-Phase Ejectors." *Inventions* 4 (15): 1–73. https://doi.org/10.3390/inventions4010015.

Aidoun, Z., K. Ameur, M. Falsafioon, and M. Badache. 2019b. "Current Advances in Ejector Modeling, Experimentation and Applications for Refrigeration and Heat Pumps. Part 2: Two-Phase Ejectors." *Inventions* 4 (16): 1–54. https://doi.org/10.3390/inventions4010016.

Akteriana, Stepan. 2011. "Improving the Energy Efficiency of Traditional Multi-Stage Steam-Jet-Ejector Vacuum Systems for Deodorizing Edible Oils." *Procedia Food Science* 1 (Icef 11): 1785–91. https://doi.org/10.1016/j.profoo.2011.09.262.

Al-ansary, Hany A. M. 2007. "The Use of Ejector Refrigeration Systems for Turbine Inlet Air Cooling: A Thermodynamic and CFD Study." In *Energy Sustainability Conference*. California.

Al-Khalidy, N. 1997. "Performance of Solar Refrigeration Ejector Refrigerating Machine." In *ASHRAE Trans. Vol.103, Part 1*, 56–64.

Alexis, G. K., and E. K. Karayiannis. 2005. "A Solar Ejector Cooling System Using Refrigerant R134a in the Athens Area." *Renewable Energy* 30 (9): 1457–69. https://doi.org/10.1016/j.renene.2004.11.004.

Alexis, G. K., and E. D. Rogdakis. 2002. "Performance Characteristics of Two Combined Ejector-Absorption Cycles." *Applied Thermal*

*Engineering* 22 (1): 97–106. https://doi.org/10.1016/S1359-4311(01)00057-6.

Aphornratana, S., S. Chungpaibulpatana, and P. Srikhirin. 2001. "Experimental Investigation of an Ejector Refrigerator: Effect of Mixing Chamber Geometry on System Performance." *International Journal of Energy Research* 25 (5): 397–411.

Aphornratana, Satha, and Ian W. Eames. 1998. "Experimental Investigation of a Combined Ejector-absorption Refrigerator." *International Journal of Energy Research* 22 (3): 198–207.

ASHRAE. 1965. "Steam-Jet Refrigeration Equipment." In *Guide and Data Book*, 655–60.

ASHRAE. 1983. "Steam-Jet Refrigeration Equipment." In *ASHRAE Equipment Handbook*, 13.1-13.6. Atlanta, USA.

Bai, Tao, Gang Yan, and Jianlin Yu. 2018. "Experimental Investigation of an Ejector-Enhanced Auto-Cascade Refrigeration System." *Applied Thermal Engineering* 129: 792–801. https://doi.org/10.1016/j.applthermaleng.2017.10.053.

Banasiak, Krzysztof, Armin Hafner, Ekaterini E. Kriezi, Kenneth B. Madsen, Michael Birkelund, Kristian Fredslund, and Rickard Olsson. 2015. "Development and Performance Mapping of a Multi-Ejector Expansion Work Recovery Pack for R744 Vapour Compression Units." *International Journal of Refrigeration* 57: 265–76. https://doi.org/10.1016/j.ijrefrig.2015.05.016.

Besagni, Giorgio. 2019. "Ejectors on the Cutting Edge: The Past, the Present and the Perspective." *Energy* 170: 998–1003. https://doi.org/10.1016/J.ENERGY.2018.12.214.

Besagni, Giorgio, Riccardo Mereu, and Fabio Inzoli. 2016. "Ejector Refrigeration: A Comprehensive Review." *Renewable and Sustainable Energy Reviews* 53: 373–407. https://doi.org/10.1016/j.rser.2015.08.059.

Besagni, Giorgio, Riccardo Mereu, Fabio Inzoli, and Paolo Chiesa. 2017. "Application of an Integrated Lumped Parameter-CFD Approach to Evaluate the Ejector-Driven Anode Recirculation in a PEM Fuel Cell

System." *Applied Thermal Engineering* 121: 628–51. https://doi.org/10.1016/j.applthermaleng.2017.04.111.

Boccardi, G., F. Botticella, G. Lillo, R. Mastrullo, A. W. Mauro, and R. Trinchieri. 2017. "Experimental Investigation on the Performance of a Transcritical $CO_2$ Heat Pump with Multi-Ejector Expansion System." *International Journal of Refrigeration* 82: 389–400. https://doi.org/10.1016/j.ijrefrig.2017.06.013.

Bowrey, R. G., V. B. Dang, and G. D. Sergeant. 1986. "An Energy Model to Minimize Energy Consumption in a Low- Temperature Operation, Steam Ejector-Cooling System." *Journal of the Energy Institute* 45: 45–48.

Boyaghchi, Fateme Ahmadi, and Parisa Heidarnejad. 2015. "Thermoeconomic Assessment and Multi Objective Optimization of a Solar Micro CCHP Based on Organic Rankine Cycle for Domestic Application." *Energy Conversion and Management* 97: 224–34. https://doi.org/10.1016/j.enconman.2015.03.036.

Butrymowicz, Dariusz, Kamil Śmierciew, and Jarosław Karwacki. 2014. "Investigation of Internal Heat Transfer in Ejection Refrigeration Systems." *International Journal of Refrigeration* 40: 131–39. https://doi.org/10.1016/j.ijrefrig.2013.11.016.

Butrymowicz, Dariusz, Kamil Śmierciew, Jarosław Karwacki, and Jerzy Gagan. 2014. "Experimental Investigations of Low-Temperature Driven Ejection Refrigeration Cycle Operating with Isobutane." *International Journal of Refrigeration* 39: 196–209. https://doi.org/10.1016/j.ijrefrig.2013.10.008.

Chandra, Vineet V., and M. R. Ahmed. 2014. "Experimental and Computational Studies on a Steam Jet Refrigeration System with Constant Area and Variable Area Ejectors." *Energy Conversion and Management* 79: 377–86. https://doi.org/10.1016/j.enconman.2013.12.035.

Chen, Guangming, Rongtao Zhang, Daibin Zhu, Shaojie Chen, Lingyun Fang, and Xinyue Hao. 2017. "Experimental Study on Two-Stage Ejector Refrigeration System Driven by Two Heat Sources."

*International Journal of Refrigeration* 74: 293–301. https://doi.org/10.1016/j.ijrefrig.2016.10.010.

Chen, Jiaheng, Jianlin Yu, and Suxin Qian. 2019. "Subcooling Control Method for the Adjustable Ejector in the Direct Expansion Solar Assisted Ejector-Compression Heat Pump Water Heater." *Applied Thermal Engineering* 148 (August 2018): 662–73. https://doi.org/10.1016/j.applthermaleng.2018.11.086.

Chen, Jinwei, Yao Chen, Huisheng Zhang, and Shilie Weng. 2018. "Effect of Different Operating Strategies for a SOFC-GT Hybrid System Equipped with Anode and Cathode Ejectors." *Energy* 163: 1–14. https://doi.org/10.1016/j.energy.2018.08.032.

Chen, Jinwei, Jingxuan Li, Dengji Zhou, Huisheng Zhang, and Shilie Weng. 2018. "Control Strategy Design for a SOFC-GT Hybrid System Equipped with Anode and Cathode Recirculation Ejectors." *Applied Thermal Engineering* 132: 67–79. https://doi.org/10.1016/j.applthermaleng.2017.12.079.

Chen, Li Ting. 1978. "A Heat Driven Mobile Refrigeration Cycle Analysis." *Energy Conversion* 18 (1): 25–29. https://doi.org/10.1016/0013-7480(78)90085-2.

Chen, Shaojie, Guangming Chen, and Lingyun Fang. 2015. "An Experimental Study and 1-D Analysis of an Ejector with a Movable Primary Nozzle That Operates with R236fa." *International Journal of Refrigeration* 60: 19–25. https://doi.org/10.1016/j.ijrefrig.2015.08.011.

Chen, Xiangjie, Mark Worall, Siddig Omer, Yuehong Su, and Saffa Riffat. 2013. "Theoretical Studies of a Hybrid Ejector CO2 Compression Cooling System for Vehicles and Preliminary Experimental Investigations of an Ejector Cycle." *Applied Energy* 102: 931–42. https://doi.org/10.1016/j.apenergy.2012.09.032.

Choutapalli, Isaac M., Krothapalli Anjaneyulu, and Mehmet B. Alkislar. 2012. "Pulsed-Jet Ejector Thrust Augumentor Characteristics." *Journal of Propulsion and Power* 28 (2): 293–306.

Dai, Yiping, Jiangfeng Wang, and Lin Gao. 2009. "Exergy Analysis, Parametric Analysis and Optimization for a Novel Combined Power and

Ejector Refrigeration Cycle." *Applied Thermal Engineering* 29 (10): 1983–90. https://doi.org/10.1016/j.applthermaleng.2008.09.016.

Daubert-Deleris, I., P. A. Hoffmann, C. Fonade, and C. Maranges. 2006. "Hydrodynamic and Mass Transfer Performance of a New Aero-Ejector with Its Application to VOC Abatement." *Chemical Engineering Science* 61 (15): 4982–93. https://doi.org/10.1016/j.ces.2006.03.023.

Eames, I. W., V. Petrenko, and A. Ablwaifa. 2004. "Design and Experimental Investigation of a Jet Pump Refrigerator." In *HPC 2004 3rd Intl. Conf. on Heat Powered Cycles*. Cyprus.

Eames, I. W., S Aphornratana, and H. Haider. 1995. "A Theoretical and Experimental Study of a Small-Scale Steam Jet Refrigerator." *International Journal of Refrigeration* 18 (6): 378–86. https://doi.org/10.1016/0140-7007(95)98160-M.

Eames, Ian W., and Shenyi Wu. 2000. "Experimental Proof-of-Concept Testing of an Innovative Heat-Powered Vapour Recompression-Absorption Refrigerator Cycle." *Applied Thermal Engineering* 20 (8): 721–36. https://doi.org/10.1016/S1359-4311(99)00054-X.

Ebrahimi, Masood, Ali Keshavarz, and Arash Jamali. 2012. "Energy and Exergy Analyses of a Micro-Steam CCHP Cycle for a Residential Building." *Energy and Buildings* 45: 202–10. https://doi.org/10.1016/j.enbuild.2011.11.009.

Fahmy, A., D. Mewes, and K. Ohlrogge. 2002. "On the Integration of Jet Ejectors into Hybrid Dehydration Processes." *Journal of Membrane Science* 196 (1): 79–84.

Fan, Chaochao, Gang Yan, and Jianlin Yu. 2019. "Thermodynamic Analysis of a Modified Solar Assisted Ejector-Compression Heat Pump Cycle with Zeotropic Mixture R290/R600a." *Applied Thermal Engineering* 150 (August 2018): 42–49. https://doi.org/10.1016/j.applthermaleng.2019.01.011.

Fang, Yu, Sergio Croquer, Sébastien Poncet, Zine Aidoun, and Yann Bartosiewicz. 2017. "Drop-in Replacement in a R134 Ejector Refrigeration Cycle by HFO Refrigerants." *International Journal of Refrigeration* 77: 87–98. https://doi.org/10.1016/j.ijrefrig.2017.02.028.

Ferrari, M. L., M. Pascenti, and A. F. Massardo. 2018. “Validated Ejector Model for Hybrid System Applications.” *Energy* 162: 1106–14. https://doi.org/10.1016/j.energy.2018.08.087.

Gagan, Jerzy, Kamil Śmierciew, and Dariusz Butrymowicz. 2018. “Performance of Ejection Refrigeration System Operating with R-1234ze(E) Driven by Ultra-Low Grade Heat Source.” *International Journal of Refrigeration* 88: 458–71. https://doi.org/10.1016/j.ijrefrig.2018.02.014.

García Del Valle, J., J. M. Saíz Jabardo, F. Castro Ruiz, and J. F. San José Alonso. 2014. “An Experimental Investigation of a R-134a Ejector Refrigeration System.” *International Journal of Refrigeration* 46: 105–13. https://doi.org/10.1016/j.ijrefrig.2014.05.028.

Ghaebi, Hadi, Towhid Parikhani, Hadi Rostamzadeh, and Behzad Farhang. 2018. “Proposal and Assessment of a Novel Geothermal Combined Cooling and Power Cycle Based on Kalina and Ejector Refrigeration Cycles.” *Applied Thermal Engineering* 130: 767–81. https://doi.org/10.1016/j.applthermaleng.2017.11.067.

Gil, Bartosz, and Jacek Kasperski. 2018. “Efficiency Evaluation of the Ejector Cooling Cycle Using a New Generation of HFO/HCFO Refrigerant as a R134a Replacement.” *Energies* 11 (8): 1–18. https://doi.org/10.3390/en11082136.

Habibzadeh, A., M. M. Rashidi, and N. Galanis. 2013. “Analysis of a Combined Power and Ejector-Refrigeration Cycle Using Low Temperature Heat.” *Energy Conversion and Management* 65: 381–91. https://doi.org/10.1016/j.enconman.2012.08.020.

Haghparast, Payam, Mikhail V. Sorin, Marc A. Richard, and Hakim Nesreddine. 2019. “Analysis and Design Optimization of an Ejector Integrated into an Organic Rankine Cycle.” *Applied Thermal Engineering* 159 (June): 113979. https://doi.org/10.1016/j.applthermaleng.2019.113979.

Haida, Michal, Krzysztof Banasiak, Jacek Smolka, Armin Hafner, and Trygve M. Eikevik. 2016. “Experimental Analysis of the R744 Vapour Compression Rack Equipped with the Multi-Ejector Expansion Work

Recovery Module." *International Journal of Refrigeration* 64: 93–107. https://doi.org/10.1016/j.ijrefrig.2016.01.017.

Hamzaoui, Malek, Hakim Nesreddine, Zine Aidoun, and Mourad Balistrou. 2018. "Experimental Study of a Low Grade Heat Driven Ejector Cooling System Using the Working Fluid R245fa." *International Journal of Refrigeration* 86: 388–400. https://doi.org/10.1016/j.ijrefrig.2017.11.018.

Hao, Xinyue, Lin Wang, Zhanwei Wang, Yingying Tan, and Xiaona Yan. 2018. "Hybrid Auto-Cascade Refrigeration System Coupled with a Heat-Driven Ejector Cooling Cycle." *Energy* 161: 988–98. https://doi.org/10.1016/j.energy.2018.07.201.

He, S., Y. Li, and R. Z. Wang. 2009. "A New Approach to Performance Analysis of Ejector Refrigeration System Using Grey System Theory." *Applied Thermal Engineering* 29 (8–9): 1592–97. https://doi.org/10.1016/j.applthermaleng.2008.07.016.

He, Yang, Jianqiang Deng, Lixing Zheng, and Zaoxiao Zhang. 2017. "Performance Optimization of a Transcritical CO2 Refrigeration System Using a Controlled Ejector." *International Journal of Refrigeration* 75: 250–61. https://doi.org/10.1016/j.ijrefrig.2016.12.015.

Hernandez, Jorge I., Ruben J. Dorantes, Roberto Best, and Claudio A. Estrada. 2004. "The Behaviour of a Hybrid Compressor and Ejector Refrigeration System with Refrigerants 134a and 142b." *Applied Thermal Engineering* 24 (13): 1765–83. https://doi.org/10.1016/j.applthermaleng.2003.12.016.

Hosseinzadeh, Elham, Masoud Rokni, Masoud Jabbari, and Henrik Mortensen. 2014. "Numerical Analysis of Transport Phenomena for Designing of Ejector in PEM Forklift System." *International Journal of Hydrogen Energy* 39 (12): 6664–74. https://doi.org/10.1016/j.ijhydene.2014.02.061.

Huang, B. J., C. B. Jiang, and F. L. Hu. 1985. "Ejector Performance Characteristics and Design Analysis of Jet Refrigeration System." *Journal of Engineering for Gas Turbine and Power* 107: 792–802.

Hwang, Jenn Jiang. 2014. "Passive Hydrogen Recovery Schemes Using a Vacuum Ejector in a Proton Exchange Membrane Fuel Cell System." *Journal of Power Sources* 247: 256–63. https://doi.org/10.1016/j.jpowsour.2013.08.126.

Jaruwongwittaya, Tawatchai, and Guangming Chen. 2012. "Application of Two Stage Ejector Cooling System in a Bus." *Energy Procedia* 14: 187–97. https://doi.org/10.1016/j.egypro.2011.12.916.

Jeon, Yongseok, Dongwoo Kim, Jongho Jung, Dong Soo Jang, and Yongchan Kim. 2018. "Comparative Performance Evaluation of Conventional and Condenser Outlet Split Ejector-Based Domestic Refrigerator-Freezers Using R600a." *Energy* 161: 1085–95. https://doi.org/10.1016/j.energy.2018.08.007.

Kasperski, Jacek, and Bartosz Gil. 2014. "Performance Estimation of Ejector Cycles Using Heavier Hydrocarbon Refrigerants." *Applied Thermal Engineering* 71 (1): 197–203. https://doi.org/10.1016/j.applthermaleng.2014.06.057.

Khaliq, Abdul, Rajesh Kumar, and Esmail M.A. Mokheimer. 2018. "Investigation on a Solar Thermal Power and Ejector-Absorption Refrigeration System Based on First and Second Law Analyses." *Energy* 164: 1030–43. https://doi.org/10.1016/j.energy.2018.09.049.

Kim, Sunjae, Yongseok Jeon, Hyun Joon Chung, and Yongchan Kim. 2018. "Performance Optimization of an R410A Air-Conditioner with a Dual Evaporator Ejector Cycle Based on Cooling Seasonal Performance Factor." *Applied Thermal Engineering, Accepted Manuscript* 131: 988–97. https://doi.org/10.1016/j.applthermaleng.2017.12.012.

Kitrattana, Borirak, Satha Aphornratana, Tongchana Thongtip, and Natthawut Ruangtrakoon. 2017. "Comparison of Traditional and CRMC Ejector Performance Used in a Steam Ejector Refrigeration." *Energy Procedia* 138: 476–81. https://doi.org/10.1016/j.egypro.2017.10.229.

Kumar, Anil, and Raj Kumar. 2014. "Thermodynamic Analysis of a Novel Compact Power Generation and Waste Heat Operated Absorption, Ejector-Jet Pump Refrigeration Cycle." *Journal of Mechanical Science and Technology* 28 (9): 3895–3902. https://doi.org/10.1007/s12206-014-0340-y.

Lawrence, Neal, and Stefan Elbel. 2016. "Experimental Investigation on the Effect of Evaporator Design and Application of Work Recovery on the Performance of Two-Phase Ejector Liquid Recirculation Cycles with R410A." *Applied Thermal Engineering* 100: 398–411. https://doi.org/10.1016/j.applthermaleng.2016.01.079.

Li, Chennan, D. Yogi Goswami, Andrew Shapiro, Elias K. Stefanakos, and Gokmen Demirkaya. 2012. "A New Combined Power and Desalination System Driven by Low Grade Heat for Concentrated Brine." *Energy* 46 (1): 582–95. https://doi.org/10.1016/j.energy.2012.07.050.

Li, Fenglei, Zhao Chang, Xinchang Li, and Qi Tian. 2018. "Energy and Exergy Analyses of a Solar-Driven Ejector-Cascade Heat Pump Cycle." *Energy* 165: 419–31. https://doi.org/10.1016/j.energy.2018.09.173.

Li, Huashan, Fei Cao, Xianbiao Bu, Lingbao Wang, and Xianlong Wang. 2014. "Performance Characteristics of R1234yf Ejector-Expansion Refrigeration Cycle." *Applied Energy* 121 (May): 96–103. https://doi.org/10.1016/j.apenergy.2014.01.079.

Li, Xinguo, Xiajie Li, and Qilin Zhang. 2013. "The First and Second Law Analysis on an Organic Rankine Cycle with Ejector." *Solar Energy* 93: 100–108. https://doi.org/10.1016/j.solener.2013.04.003.

Li, Xinguo, Qilin Zhang, and Xiajie Li. 2013. "A Kalina Cycle with Ejector." *Energy* 54: 212–19. https://doi.org/10.1016/j.energy.2013.03.040.

Li, Zishen, Weiyi Li, and Borui Xu. 2016. "Optimization of Mixed Working Fluids for a Novel Trigeneration System Based on Organic Rankine Cycle Installed with Heat Pumps." *Applied Thermal Engineering* 94: 754–62. https://doi.org/10.1016/j.applthermaleng.2015.10.145.

Lijo, Vincent, Heuy Dong Kim, Shigeru Matsuo, and Toshiaki Setoguchi. 2012. "A Study of the Supersonic Ejector-Diffuser System with an Inlet Orifice." *Aerospace Science and Technology* 23 (1): 321–29. https://doi.org/10.1016/j.ast.2011.08.009.

Liu, Ye, Jianlin Yu, and Gang Yan. 2018. "Theoretical Analysis of a Double Ejector-Expansion Autocascade Refrigeration Cycle Using Hydrocarbon Mixture R290/R170." *International Journal of Refrigeration* 94: 33–39. https://doi.org/10.1016/j.ijrefrig.2018.07.025.

Maghsoodi, A., E. Afshari, and H. Ahmadikia. 2014. "Optimization of Geometric Parameters for Design a High-Performance Ejector in the Proton Exchange Membrane Fuel Cell System Using Artificial Neural Network and Genetic Algorithm." *Applied Thermal Engineering* 71 (1).

Man-O, T., M. Tanino, T. Okazaki, and S. Koyama. 2007. "The Enhanced Heat Transfer in the Plate-Type Evaporator by Using an Ejector for Recirculation." In *International Congress of Refrigeration.* Beijing, China. ICR07-B2-931.

Mazzelli, Federico, and Adriano Milazzo. 2015. "Performance Analysis of a Supersonic Ejector Cycle Working with R245fa." *International Journal of Refrigeration* 49 (0): 79–92.

Megdouli, K., B. M. Tashtoush, E. Nahdi, M. Elakhdar, L. Kairouani, and A. Mhimid. 2016. "Thermodynamic Analysis of a Novel Ejector-Cascade Refrigeration Cycles for Freezing Process Applications and Air-Conditioning." *International Journal of Refrigeration* 70: 108–18. https://doi.org/10.1016/j.ijrefrig.2016.06.029.

Mohamed, Saleh, Youssef Shatilla, and Tiejun Zhang. 2019. "CFD-Based Design and Simulation of Hydrocarbon Ejector for Cooling." *Energy* 167: 346–58. https://doi.org/10.1016/j.energy.2018.10.057.

Mondal, Subha, Shahbaz Alam, and Sudipta De. 2018. "Performance Assessment of a Low Grade Waste Heat Driven Organic Flash Cycle (OFC) with Ejector." *Energy* 163: 849–62. https://doi.org/10.1016/j.energy.2018.08.160.

Nesreddine, Hakim, Adlane Bendaoud, Zine Aidoun, and Mohamed Ouzzane. 2015. "Experimental Investigation of an Ejector-Compression." In *ICR 2015*, 1051–54. Yokohama, Japan. ID 569.

Oliveira, A. C., C. Afonso, J. Matos, S. Riffat, M. Nguyen, and P. Doherty. 2002. "A Combined Heat and Power System for Buildings Driven by Solar Energy and Gas." *Applied Thermal Engineering* 22 (6): 587–93. https://doi.org/10.1016/S1359-4311(01)00110-7.

Padilla, R. V., Y. C. S. Too, R. Benito, R. Mcnaughton, and W. Stein. 2016. "Thermodynamic Feasibility of Alternative Supercritical CO2 Brayton Cycles Integrated with an Ejector." *Applied Energy* 169: 49–62. https://doi.org/10.1016/j.apenergy.2016.02.029.

Palacz, Michal, Michal Haida, Jacek Smolka, Marcin Plis, Andrzej J. Nowak, and Krzysztof Banasiak. 2018. "A Gas Ejector for CO2 Supercritical Cycles." *Energy* 163: 1207–16. https://doi.org/10.1016/j.energy.2018.09.030.

Pereira, Paulo R., Szabolcs Varga, João Soares, Armando C. Oliveira, António M. Lopes, Fernando G. De Almeida, and João F. Carneiro. 2014. "Experimental Results with a Variable Geometry Ejector Using R600a as Working Fluid." *International Journal of Refrigeration* 46: 77–85. https://doi.org/10.1016/j.ijrefrig.2014.06.016.

Petrenko, V. 2009. "Application of Innovative Ejector Chillers and Air Conditioners Operating with Low Boiling Refrigerants in Tri-Generation Systems." In *Int. Sem. on Ejectors*. Louvain, Belgium.

Petrenko, V. O., B. J. Huang, and V. O. Ierin. 2011. "Design-Theoretical Study of Cascade CO2 Sub-Critical Mechanical Compression/Butane Ejector Cooling Cycle." *International Journal of Refrigeration* 34 (7): 1649–56. https://doi.org/10.1016/j.ijrefrig.2010.11.012.

Popovac, Mirza, Michael Lauermann, Alexander Baumhakel, and Gerwin Drexler-. 2017. "Performance Analysis of a High-Temperature Heat Pump with Ejector Based on Butane as the Refrigerant." *12th IEA Heat Pump Conference 2017, Rotterdam*, 1–8.

Purohit, Nilesh, Vishaldeep Sharma, Samer Sawalha, Brian Fricke, Rodrigo Llopis, and Mani Sankar Dasgupta. 2018. "Integrated Supermarket Refrigeration for Very High Ambient Temperature." *Energy* 165: 572–90. https://doi.org/10.1016/j.energy.2018.09.097.

Rao, S. M. V., and G. Jagadeesh. 2015. "Studies on the Effects of Varying Secondary Gas Properties in a Low Entrainment Ratio Supersonic Ejector." *Applied Thermal Engineering* 78: 289–302. https://doi.org/10.1016/j.applthermaleng.2014.12.053.

Reddick, Christopher, Mikhail Sorin, Hristo Sapoundjiev, and Zine Aidoun. 2016. "Carbon Capture Simulation Using Ejectors for Waste Heat Upgrading." *Energy* 100: 251–61. https://doi.org/10.1016/j.energy.2016.01.099.

Reddick, Christopher, Mikhail Sorin, Hristo Sapoundjiev, and Zine Aidoun. 2018. "Effect of a Mixture of Carbon Dioxide and Steam on Ejector

Performance: An Experimental Parametric Investigation." *Experimental Thermal and Fluid Science* 92: 353–65.

Ruangtrakoon, Natthawut, and Satha Aphornratana. 2014. "Development and Performance of Steam Ejector Refrigeration System Operated in Real Application in Thailand." *International Journal of Refrigeration* 48: 142–52. https://doi.org/10.1016/j.ijrefrig.2014.09.010.

Salim, M. 2004. "Thermally Activated Mobile Ejector Refrigeration System Analysis." In *Proceedings of the Institution of Mechanical Engineers, Part D: Journal of Automobile Engineering.*, 218:1055–61.

Santini, Fabrizio, Giuseppe Bianchi, Davide Di Battista, Carlo Villante, and Maurizio Orlandi. 2019. "Experimental Investigations on a Transcritical $CO_2$ Refrigeration Plant and Theoretical Comparison with an Ejector-Based One." *Energy Procedia* 161: 309–16. https://doi.org/10.1016/j.egypro.2019.02.097.

Selvaraju, A., and A. Mani. 2006. "Experimental Investigation on R134a Vapour Ejector Refrigeration System." *International Journal of Refrigeration* 29 (7): 1160–66. https://doi.org/10.1016/j.ijrefrig.2006.01.004.

Shan, Yong, Jing zhou Zhang, and Xiao wen Ren. 2018. "Numerical Modeling on Pumping Performance of Piccolo-Tube Multi-Nozzles Supersonic Ejector in an Oil Radiator Passage." *Energy* 158: 216–27. https://doi.org/10.1016/j.energy.2018.06.001.

Shaozhi, Z., J. Luo, Q. Wang, and G. Chen. 2018. "Step Utilization of Energy with Ejector in a Heat Driven Freeze Drying System." *Energy* 164: 734–44. https://doi.org/10.1016/j.energy.2018.08.195.

Shestopalov, K. O., B. J. Huang, V. O. Petrenko, and O. S. Volovyk. 2015. "Investigation of an Experimental Ejector Refrigeration Machine Operating with Refrigerant R245fa at Design and Off-Design Working Conditions. Part 1. Theoretical Analysis." *International Journal of Refrigeration* 55: 201–11. https://doi.org/10.1016/j.ijrefrig.2015.01.016.

Shi, Yuqi, Fei Li, Daliang Hong, Qin Wang, and Guangming Chen. 2018. "Experimental Study of a New Ejector-Absorption Refrigeration Cycle

Driven by Multi-Heat Sources." *Applied Thermal Engineering* 133: 604–12. https://doi.org/10.1016/j.applthermaleng.2018.01.073.

Sirwan, Ranj, M. A. Alghoul, K. Sopian, Yusoff Ali, and Jasim Abdulateef. 2013. "Evaluation of Adding Flash Tank to Solar Combined Ejector-Absorption Refrigeration System." *Solar Energy* 91: 283–96. https://doi.org/10.1016/j.solener.2013.01.018.

Śmierciew, Kamil, Jerzy Gagan, Dariusz Butrymowicz, Michał Łukaszuk, and Henryk Kubiczek. 2017. "Experimental Investigation of the First Prototype Ejector Refrigeration System with HFO-1234ze(E)." *Applied Thermal Engineering* 110: 115–25. https://doi.org/10.1016/j.applthermaleng.2016.08.140.

Sun, D. W., and I. W. Eames. 1996. "Performance Characteristics of HCFC-123 Ejector Refrigeration Cycles." *International Journal of Energy Research* 20 (10): 871–85.

Sun, Da-Wen, Ian W. Eames, and Satha Aphornratana. 1996. "Evaluation of a Novel Combined Ejector-Absorption Refrigeration Cycle — I: Computer Simulation." *International Journal of Refrigeration* 19 (3): 172–80. https://doi.org/10.1016/0140-7007(96)00010-2.

Sun, Fangtian, Xu Chen, Lin Fu, and Shigang Zhang. 2018. "Configuration Optimization of an Enhanced Ejector Heat Exchanger Based on an Ejector Refrigerator and a Plate Heat Exchanger." *Energy* 164: 408–17. https://doi.org/10.1016/j.energy.2018.08.194.

Sun, Fangtian, Lin Fu, Jian Sun, and Shigang Zhang. 2014. "A New Waste Heat District Heating System with Combined Heat and Power (CHP) Based on Ejector Heat Exchangers and Absorption Heat Pumps." *Energy* 69: 516–24. https://doi.org/10.1016/j.energy.2014.03.044.

Takleh, H. R., and V. Zare. 2019. "Performance Improvement of Ejector Expansion Refrigeration Cycles Employing a Booster Compressor Using Different Refrigerants: Thermodynamic Analysis and Optimization." *International Journal of Refrigeration* 101: 56–70. https://doi.org/10.1016/j.ijrefrig.2019.02.031.

Tang, Yongzhi, Zhongliang Liu, Can Shi, and Yanxia Li. 2018a. "A Novel Steam Ejector with Pressure Regulation to Dredge the Blocked Entrained Flow for Performance Improvement in MED-TVC

Desalination System." *Energy Conversion and Management* 172 (June): 237–47. https://doi.org/10.1016/j.enconman.2018.07.022.

Tang, Yongzhi, Zhongliang Liu, Can Shi, and Yanxia Li. 2018b. "A Novel Steam Ejector with Pressure Regulation to Optimize the Entrained Flow Passage for Performance Improvement in MED-TVC Desalination System." *Energy* 158: 305–16. https://doi.org/10.1016/j.energy.2018.06.028.

Tashtoush, B. M., M. A. Al-Nimr, and M. A. Khasawneh. 2019. "A Comprehensive Review of Ejector Design, Performance, and Applications." *Applied Energy* 240 (February): 138–72. https://doi.org/10.1016/j.apenergy.2019.01.185.

Thongtip, Tongchana, and Satha Aphornratana. 2015. "An Alternative Analysis Applied to Investigate the Ejector Performance Used in R141b Jet-Pump Refrigeration System." *International Journal of Refrigeration* 53: 20–33. https://doi.org/10.1016/j.ijrefrig.2015.01.017.

Thongtip, Tongchana, and Satha Aphornratana. 2017. "An Experimental Analysis of the Impact of Primary Nozzle Geometries on the Ejector Performance Used in R141b Ejector Refrigerator." *Applied Thermal Engineering* 110: 89–101. https://doi.org/10.1016/j.applthermaleng.2016.08.100.

Wang, Lei, Wenjian Cai, Hongxia Zhao, Chen Lin, and Jia Yan. 2016. "Experimentation and Cycle Performance Prediction of Hybrid A/C System Using Automobile Exhaust Waste Heat." *Applied Thermal Engineering* 94: 314–23. https://doi.org/10.1016/j.applthermaleng.2015.10.051.

Wu, S, and Ian W. Eames. 1998. "A Novel Absorption-Recompression Refrigeration Cycle." *Applied Thermal Engineering* 18: 1149–57.

Xia, Jiaxi, Jiangfeng Wang, Kehan Zhou, Pan Zhao, and Yiping Dai. 2018. "Thermodynamic and Economic Analysis and Multi-Objective Optimization of a Novel Transcritical CO2 Rankine Cycle with an Ejector Driven by Low Grade Heat Source." *Energy* 161: 337–51. https://doi.org/10.1016/j.energy.2018.07.161.

Xu, Yingjie, Ning Jiang, Qin Wang, Ning Mao, Guangming Chen, and Zengliang Gao. 2018. "Refrigerant Evaluation and Performance

Comparison for a Novel Hybrid Solar-Assisted Ejection-Compression Refrigeration Cycle." *Solar Energy* 160: 344–52. https://doi.org/10.1016/j.solener.2017.12.030.

Yan, Jia, Wenjian Cai, Lei Zhao, Yanzhong Li, and Chen Lin. 2013. "Performance Evaluation of a Combined Ejector-Vapor Compression Cycle." *Renewable Energy* 55: 331–37. https://doi.org/10.1016/j.renene.2012.12.029.

Yan, Jiwei, Guangming Chen, Chengyan Liu, Liming Tang, and Qi Chen. 2017. "Experimental Investigations on a R134a Ejector Applied in a Refrigeration System." *Applied Thermal Engineering* 110: 1061–65. https://doi.org/10.1016/j.applthermaleng.2016.09.046.

Yang, Xingyang, Nan Zheng, Li Zhao, Shuai Deng, Hailong Li, and Zhixin Yu. 2016. "Analysis of a Novel Combined Power and Ejector-Refrigeration Cycle." *Energy Conversion and Management* 108: 266–74. https://doi.org/10.1016/j.enconman.2015.11.019.

Yapici, R., H. K. Ersoy, A. Aktoprakoğlu, H. S. Halkaci, and O. Yiğit. 2008. "Experimental Determination of the Optimum Performance of Ejector Refrigeration System Depending on Ejector Area Ratio." *International Journal of Refrigeration* 31 (7): 1183–89. https://doi.org/10.1016/j.ijrefrig.2008.02.010.

Yapici, R., and C. C. Yetişen. 2007. "Experimental Study on Ejector Refrigeration System Powered by Low Grade Heat." *Energy Conversion and Management* 48 (5): 1560–68. https://doi.org/10.1016/j.enconman.2006.11.015.

Yosaf, Salem, and Hasan Ozcan. 2018. "Exergoeconomic Investigation of Flue Gas Driven Ejector Absorption Power System Integrated with PEM Electrolyser for Hydrogen Generation." *Energy* 163: 88–99. https://doi.org/10.1016/j.energy.2018.08.033.

Zhang, Kun, Xue Chen, Christos N. Markides, Yong Yang, and Shengqiang Shen. 2016. "Evaluation of Ejector Performance for an Organic Rankine Cycle Combined Power and Cooling System." *Applied Energy* 184: 404–12. https://doi.org/10.1016/j.apenergy.2016.10.017.

Zhang, Shaozhi, Jielin Luo, Qin Wang, and Guangming Chen. 2018. "Step Utilization of Energy with Ejector in a Heat Driven Freeze Drying

System." *Energy* 164: 734–44. https://doi.org/10.1016/j.energy.2018.08.195.

Zhang, TieJun, and Saleh Mohamed. 2014. "Conceptual Design and Analysis of Hydrocarbon-Based Solar Thermal Power and Ejector Cooling Systems in Hot Climates." *Journal of Solar Energy Engineering* 137 (2): 021001. https://doi.org/10.1115/1.4028365.

Zheng, Lixing, and Jianqiang Deng. 2017. "Experimental Investigation on a Transcritical $CO_2$ Ejector Expansion Refrigeration System with Two-Stage Evaporation." *Applied Thermal Engineering* 125: 919–27. https://doi.org/10.1016/j.applthermaleng.2017.07.056.

Zhu, Yinhai, Yulei Huang, Conghui Li, Fuzhen Zhang, and Pei Xue Jiang. 2018. "Experimental Investigation on the Performance of Transcritical $CO_2$ Ejector-Expansion Heat Pump Water Heater System." *Energy Conversion and Management* 167 (March): 147–55. https://doi.org/10.1016/j.enconman.2018.04.081.

Zhu, Yinhai, Yanzhong Li, and Wenjian Cai. 2011. "Control Oriented Modeling of Ejector in Anode Gas Recirculation Solid Oxygen Fuel Cell Systems." *Energy Conversion and Management* 52 (4): 1881–89. https://doi.org/10.1016/j.enconman.2010.11.012.

In: Waste Heat: Sources and Uses
Editor: Michèle Picard
ISBN: 978-1-53617-355-0


*Chapter 3*

# WASTE HEAT RECOVERY TECHNOLOGIES IN INDUSTRIAL AND BUILDING APPLICATIONS

***Zine Aidoun[1], Messaoud Badache[1], Khaled Ameur[1] and Hakim Nesreddine[2,*]***
[1]Buildings group, CanmeENERGY, Varennes, Canada
[2]Hydro-Québec, Shawinigan, Canada

## ABSTRACT

In most industrialised countries, energy consumption is high and the demand, mainly based on fossil resources for its production, is increasing. A recent editorial document reports that industrial needs for heat amount approximately to 70% of the total demand and generate 33% of overall greenhouse gases (GHGs). In addition, up to 50% of energy input to processes end up as waste heat in hot exhaust gases, cooling water, or from equipment surfaces and heated products.

It is nowadays very clear that recovering a portio of this heat for recycling in the processes themselves or for alternative forms and uses will substantially reinforce economic and environmental benefits. For example,

* Corresponding Author's Email: nesreddine.hakim@ireq.ca.

recovered waste heat may contribute to electricity generation, preheat combustion air or boiler water as well as produce cooling and heating effects, depending on particular needs.

Efforts to use efficiently the available energy with minimal losses have thus been ongoing for a long time; research and development on related issues has been prolific. Methodologies and approaches for optimized waste heat recovery, where source-sink compatibility, temporal availability and adapted technology selection continue to be proposed. Applied research activities concern all aspects open to improvement such as waste heat characterization (type, quality, quantity, etc…), use of conventional technologies, and development of new concepts, depending on the need, the kind of process, including economic considerations.

Waste heat recovery at high temperature, at different levels, may be transferred with passive heat exchangers between streams and processes. In addition it can produce power by means of conventional steam turbines or ORCs. With decreasing temperatures, waste heat recovery becomes progressively limited in terms of energy harnessing and efficiencies. In this range of temperatures, alternative technologies and systems for temperature uplift, refrigeration and space heating are more frequently proposed for investigation.

The purpose of this chapter is to provide a concise overview of waste heat availability and energy capturing potential, as well as the different technologies available for its recovery with their advantages and challenges. In view of their importance, selected conventional and emerging technologies such as Rankine cycle, ORCs and heat pump types will be discussed in some detail. Steam ejectors are introduced in the context of specific applications where they have been used for many years and are available in various sizes and geometries. Ejector technology extended to a range of other working fluids make it suitable to harness waste heat at the lower ranges of temperature, which will receive particular attention in a separate chapter.

**Keywords**: waste heat recovery, low-grade energy, technologies for heat recovery

## 1. Introduction

In most industrialized countries, energy consumption is high and the demand, still mainly based on fossil resources for its production, is increasing. The consequences in terms of rising fuel costs and environmental concerns are therefore serious and call for urgent alleviatory measures. According to the estimations of USDOE (IEO 2019) for the period of 2015 to 2050, world energy demand will increase by 70% in non-OECD countries and by 15% in OECD countries. An editorial document reports that current industrial needs for heat amount approximately to 70% of the total demand and generate 33% of overall GHGs (Jouhara and Olabi 2018). In addition, up to 50% of the energy input to processes end up as waste heat in the form of hot exhaust gases, cooling water, or from equipment surfaces and heated products (USDOE 2008). At the same time the demand for heating and cooling exists for a wide range of applications and temperature levels: 70 to 150°C in many industrial sectors (chemical, pharmaceutical, pulp and paper, agri-food processes) and as low as 35°C in buildings (heating and cooling) as recently reported by Arpagaus et al. (2016).

In response to these challenges, efforts are made along several exploratory paths of technical developments, to satisfy future energy demands in a cost effective and sustainable manner. In this context, among the measures taken to modify the existing energy consumption schema, the following are most considered:

- the reduction of the energy intensity in industry and buildings,
- the shift from fossil fuels to electricity,
- the use of the considerable pools of industrial waste heat that may be recovered for direct use, after recompression or upgrading,
- clean power generation by means of available types of renewable energies, including solar, biomass, ground-source, etc…(Quoilin et al. 2013).

In order to fulfill the energy demand while ensuring sustainable environmental conditions, research and development on waste recovery, efficient systems and new methodologies have come to the forefront of the research activities.

# 2. Waste Heat – An Overview

The primary intent of recovering waste heat is to maximize the efficiency of systems and processes. Expected benefits are process efficiency improvement, diminished fuel consumption, reduced operating costs and greenhouse gas emissions (GHG) to environment. Substantial reductions in energy consumption are estimated to be in the order of 5 to 30% and auxiliary equipment sizes (Reddy, Naidu, and Rangaiah 2013; Mahmoudi, Fazli, and Morad 2018).

This section presents an overview of waste heat (WH) sources, categories, and recovery technologies. Challenges and barriers for waste heat recovery (WHR) are also briefly discussed.

## 2.1. Waste Heat Generation

Sources of waste heat are multiple and characterized by their abundance, nature (liquid, gas, compounds, particles, etc…), quality, thermal power and temperature range (Kolasinska and Kolasinska 2016). The main waste heat sources are the energy intensive industries such as chemical, pulp and paper, food and drink, iron and steel processes, etc…). Waste heat sources can be classified by their origins (Thekdi and Nimbalkar 2014) into the following categories (Figure 39): industrial, oil & gas processes, electricity and power systems, buildings and communities, transportation and others such as renewable energies (geothermal, solar…).

Waste heat is essentially a byproduct of industrial processes and building operation. It is this part of the consumed energy by a process that is eventually released to the environment without being converted to a useful form of energy, due to equipment inefficiencies and/or process/system thermodynamic limitations (Table1).

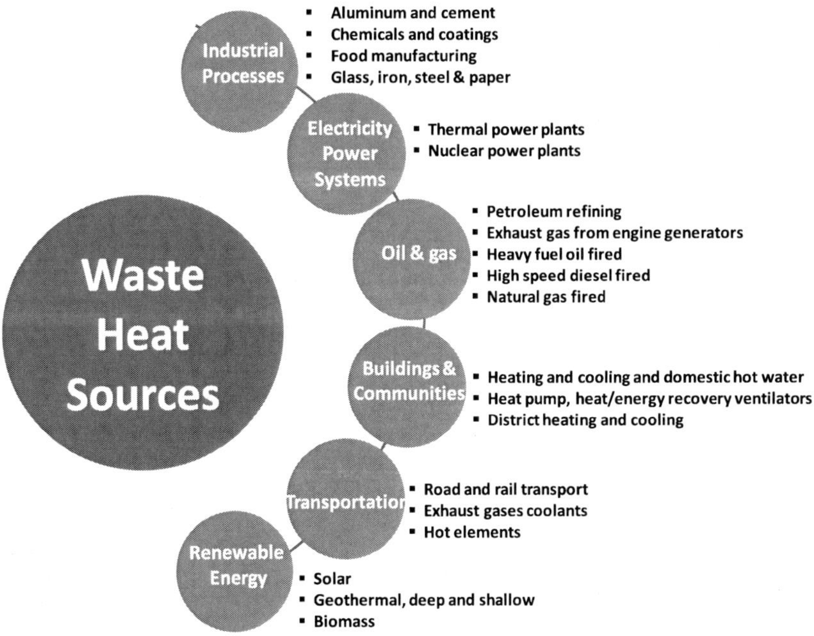


Figure 39. Sources of waste heat.

**Table 1. Typical thermal efficiencies of selected industrial processes and equipment (Pelligrino et al. 2004)**

| Equipment/Process | Efficiency (%) | Equipment/Process | Efficiency (%) |
|---|---|---|---|
| Power generation | 25-44% | Kraft pulping | 60-69% |
| Steam boilers (natural gas) | 80% | Distillation column | 25-40% |
| Steam boilers (coal and oil) | 84-85% | Cement calciner | 30-70% |
| Waste heat boilers | 60-70% | Compressors | 10-20% |
| Thermal cracking (refineries) | 58-61% | Pumps and fans | 55-65% |
| Steel making | 56% | Motors | 90-95% |
| Paper drying | 48% | | |

## 2.2. Temperature Level

The quality of waste heat depends essentially on its temperature level. Discharge temperatures vary from as low as 93°C to as high as 1650°C and several temperature categories are used in the literature (Vance et al. 2019). Typically, waste heat can be classified as USDOE (2008):

- High grade: temperature > 650°C.
- Medium grade: 230°C < temperature < 650°C.
- Low grade: temperature < 230°C.

Vance et al. (2019) have added two levels of temperature to this definition, ultra-high grade (Temperature higher than 870°C) and ultra-low temperature (Temperature below 120°C). Other classifications exist, such as the four level ranking regime of temperatures proposed by Galanis et al. (2009) and reported in Mahmoudi, Fazli, and Morad (2018) for liquid streams (50-300°C), flue gases (150-800°C), steam and process gases (100-250°C) and vapors (80-500°C).

Usually most of the waste heat in the high temperature range comes from direct combustion processes from furnaces, in the medium range from the exhaust of combustion units and in the low temperature range from parts, products and the equipment of process units (Jouhara et al. 2018).

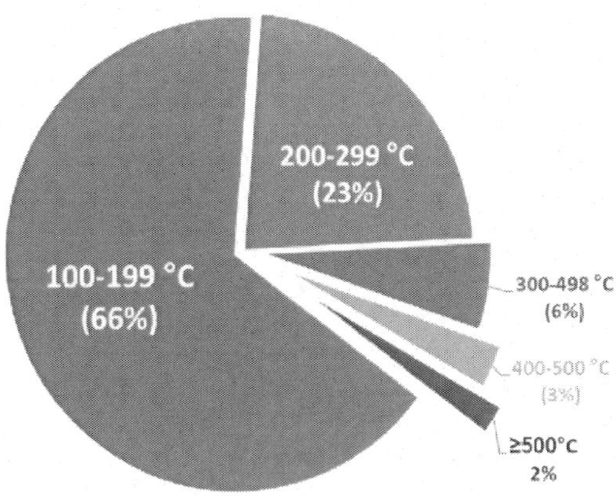


Figure 40. Waste heat temperature distribution in the UK industry (Haddad et al. 2014).

Some studies report that 20% to 50% of the energy consumed in industrial applications is finally rejected to the environment as waste heat, among which at least 60% is at low temperature (less than 230°C), and generally remains unrecovered (USDOE 2008). This trend is depicted in Figure 40, which represents waste heat temperature distribution in the UK manufacturing industry (Haddad et al. 2014).

## 2.3. End-Uses and Technologies

Recovering thermal energy for heating, cooling and generating power represents nowadays an important opportunity to improve the efficiency of processes and reduce the current overall energy consumption, operating costs and GHG emissions. Beside the use of existing recovery practices and conventional equipment to capture partially waste heat, new and different kind of technologies are being proposed, depending of the source type and quality, mostly for the low temperature range, that is particularly difficult to harness. These include a wide range of heat exchangers, heat pumps, several power and thermal cycles, heat recovery devices from exhaust gases and so on, for purposes of refrigeration, boiler feed water heating, etc.... (Agathokleous et al. 2019).

High-temperature waste heat recovery finds many uses such as combustion air preheating in furnaces or high quality steam generation in boilers for processes. In this operation, a bottoming system is typically used for process requirements and the heat rejected is further used to generate power by means of a recovery boiler and a turbine.

Medium to low temperatures typical uses include: liquid and air preheating, thermally driven technologies for heating, cooling or power production. Typical technologies used for these purposes are but not limited to Organic Rankine Cycle (ORC) and absorption cycles, ejectors, as well as steam recompression and industrial heat pumps.

Various types of natural circulating devices (heat pipes, thermosiphons and heat exchangers) may ensure passive heat recovery at all operating temperature levels.

## 2.4. Renewable Energies and WHR

Given the current energy efficiency and environmental protection context, waste heat recovery and the use of renewable energy are gaining increasing interest, which explains the amount of recent work on this subject. Probably, district heating networks (DHN) are the most representative systems combining different sources of energy (Li and Nord 2018). DHN are basically centralized thermal energy systems delivering heating or cooling to end-users through an underground network of steam or water. These networks can retrieve their heat from multiple sources: fossil fuel, biomass, solar, geothermal and waste heat from industrial processes. More details on the potential of the renewable energy solution are given in section 4.2.

## 2.5. Challenges and Barriers for WHR

Over the past decades, opportunities for waste heat recovery have not always been among the research top priorities, due essentially to:

- Low cost of energy and abundance of fossil energy resources.
- Modest energy savings expected in respect to the extent of the required effort and cost for implementing such solutions.

Over time, the effects of these factors have reduced and several strategies have shown the potential of WHR to respond to increasing energy needs while alleviating their environmental footprint.

O'Rielly and Jeswiet (2015), in their discussion of industrial waste heat harnessing potential, reported new data and a number of new case studies based on recovery practices in Canadian manufacturers' facilities. They identified some important barriers impeding their widespread implementation, such as limitations on time and resources. According to USDOE (2008), most industrial manufacturing facilities are unlikely to invest in WHR projects, if the payback period is more than three years.

Further, the lack of regulations and promotion policies has an important influence on decision-making, such that the primary focus was on measures favoring the best return on investment. Due to the fact that in some instances heat recovery is not possible or not commonly put into practice (either no use is found for it or high cost for its transfer to another site), economic considerations and perceived technical risks still remain the main obstacles.

Challenges and barriers of waste heat recovery are well addressed in the literature (Oldershaw et al. 2016; Lemmens 2015; Pelligrino et al. 2004; Thekdi and Nimbalkar 2014), and can be summarized briefly as follows:

- Economic barriers: long payback periods, maintenance and equipment costs favoring large-scale WHR systems.
- Energy price volatility and supply uncertainties.
- Regulations and safety (e.g., some regulations limit future investments in more efficient refrigerant and energy systems presenting flammable risks).
- Lack of technologies to extract and use low-grade waste heat effectively.
- Diverse industrial processes and site-specific conditions (e.g., lack of end-uses).
- Technical challenges and limitations such as durability and availability of materials for corrosive and high temperature environment; thermodynamic limitations.

It is worth recalling that while some of these barriers are specific to a given process or application, many are crosscutting across several heat recovery approaches. They generally reduce the efficiency of the adopted heat recovery systems and, in some cases, prevent the implementation of the heat recovery measure. This is mainly due to the applications' specific constraints and the nature of key restrictions (use, design, thermodynamics, cost, heat stream quality and composition, temperature, process and applications).

# 3. Waste Heat Estimation Studies

## 3.1. Methodologies and Approaches

Methodologies and approaches are proposed to systematically evaluate not only the availability of waste heat in processing and manufacturing plants, but also in various other important energy-consuming sectors. The potential and means of recovery, as well as the most appropriate technologies for the purpose of optimally harvesting and exploiting this energy are also investigated. A great deal of work has been published in this respect but only highlights of recent progress are reported in the following paragraphs. More details may be found in recent reviews by Woolley, Luo, and Simeone (2018) and Simeoni et al. (2019).

Given the diversity of types and conditions in which waste heat is generated, many studies have been performed. The aim of these studies is to develop approaches and estimate the amounts that may be recovered, where, when and how to use it and with what technology. Indeed, waste heat as already mentioned depends on the process generating it, where and how it may be applied, given its quantity, cleanliness, type and temperature level. These approaches are effective decision support tools that provide valuable guidance to exploit the recovered waste heat efficiently using available state-of-the-art technologies. However, adequate identification and use of waste heat is still somewhat tricky because the constraints are multiple and site specific (i.e., location, waste heat potential, costs, etc.), in addition to more technical considerations already mentioned (fluid nature, temperature, composition, etc…). As a result, various methodologies were proposed and their estimations are generally influenced by the weight given to the various constraints and the technologies available for heat recovery. A detailed discussion of the waste heat estimation approaches is beyond the scope of this work. The following is only highlights of the main recent approaches found in the literature.

## 3.2. Top-Down Approach

Proposed by Forman et al. (2016) to quantify global waste heat from processes, this procedure is constrained by thermodynamic efficiency limits. Energy balances are applied to individual conversion processes with estimated quality factors. The total waste heat quantity calculated is then converted into a Carnot potential. The method was applied to globally estimate the waste heat potential for the transportation, industrial, commercial and residential sectors, including electricity generation and the temperature distribution of this unused energy. The study revealed that 72% of the total input energy is lost after conversion, 63% of which arise at a temperature below 100°C with the electricity generation having the largest share before transportation and industry.

## 3.3. Optimization Approach

A novel and detailed mixed-integer linear programming optimization model based on energy cascade utilization was proposed by Xuan. Wang et al. (2018). This method is meant to use waste heat efficiently and distinguish different waste heat thermal energy qualities in distributed energy system planning and operation optimization. In this way, several energy technologies can function under different waste heat inputs to produce electricity, heating, or cooling in optimal conditions.

More recently Simeoni et al. (2019) developed a multi-objective optimization model to evaluate the sustainability of systems involving a waste heat source represented by an industrial facility and a district heating network. Application to a case study involving a heat casting facility, allowed the authors to select adequately a district heating set of users, in order to optimally recover the waste heat available. Issues like simultaneously meeting the objectives of the different actors involved, while ensuring a compromise solution between economic and environmental considerations were accounted for.

## 3.4. Framework Approach

This approach suggested by Woolley, Luo, and Simeone (2018) who worked on a framework to assist manufacturers in identifying and categorizing the waste heat generated in their facilities in four steps shown in (Figure 41). They developed a support tool for managers and decision makers allowing to analyze the compatibility between waste heat sources and sinks in terms of exergy balance and temporal availability, select adapted heat recovery technologies and estimate costs and potential economic benefits. The framework relies on historical and/or simulated data to match sources and sinks, in order to select appropriate technologies for energy recovery. A case study to demonstrate the validity of the approach and the economic viability in practice was conducted.

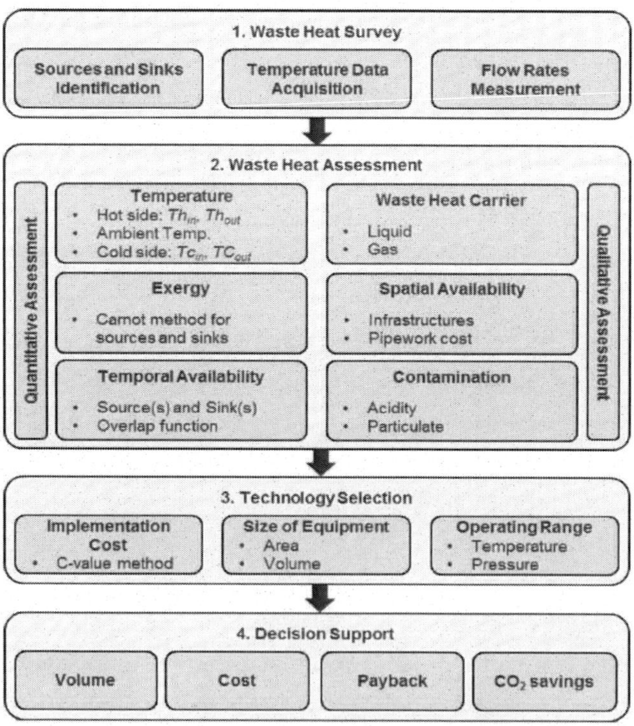


Figure 41. Framework for energy recovery assessment; (Woolley, Luo, and Simeone 2018).

## 3.5. Other Approaches

Papapetrou et al. (2018) proposed a method developed in two steps, the first relies on waste heat fractions collected in UK over four years (2000-2003). These fractions reflect the waste heat consumed and potentially recoverable. The authors computed these fractions for every main industrial sector and temperature level, adjusted to the conditions of EU countries corresponding to the same time-period, in order to account for the different levels of energy efficiency. A second adjustment was then applied in order to adjust the fractions for the year 2015, based on data for the evolution of energy intensity values up to this date. Again, the adjustment concerned each country and each sector. The main results of the analysis for EU in terms of waste heat temperature distributions were respectively 33% corresponding to temperature level below 200°C, which is often referred to as low level, 25% in the range 200-500°C and the rest above 500°C (mostly in the range 500-1000°C).

Firth, Zhang, and Yang (2019) method is meant to quantify and analyze current global waste heat generation but also extends the prediction to the future and anticipates the potential environmental impacts. Scenarios up to 2030 were produced for waste heat generation in the building, industry, transport, and power sectors.

According to Miró, Brückner, and Cabeza (2015), reliable data of industrial waste heat potential per region is the key point. By collecting information from a large sample of countries and sub-regions to build their model, the authors introduced factors such as Energy Consumed by the Country (ECC), Energy Consumed by the Industry (ECI) and amount of Industrial Waste Heat (IWH) intensive industry in the country. In their methodology two ratios per country are calculated: the $\frac{\text{IWH}}{\text{ECI}}$ and the $\frac{\text{IWH}}{\text{ECC}}$. By using these ratios, the waste heat becomes independent of the size of the region considered, and data are more comparable.

Finally, Jouhara et al. (2018) produced a detailed review of waste heat recovery and applications concerns more the practices and technologies of common use in industrial processes than the waste heat distributions or their particularities. The authors systematically assessed heat recovery opportunities and current technical features of a wide range of industries. It concerns the performance of the technologies used, their objectives and ranges of application. While at it, the main technologies across the temperature range were explored in terms of operation, advantages and limitations.

## 4. CURRENT OPPORTUNITIES OF WASTE HEAT CAPTURE

Waste heat recovery consists in capturing part of the energy input to a process that normally would not be used and rejected in the environment. Thus, if recovered, a substantial amount of energy can be reintroduced to increase process efficiency, reduce consumption, and improve the environmental footprint of products and services. The recovered heat can be delivered to any process or device that would effectively use it, to save energy, reduce costs and GHGs. For example, recovered waste heat may serve to generate electricity, preheat process fluids or provide heating and cooling.

When the temperature level matches the conditions of the intended use, the recovered heat is used directly (possibly by mixing if there are no contamination issues) or passively by means of heat exchangers or thermosiphons. Application opportunities in these conditions are more frequent in the higher temperature range. In the medium range, industrial vapor recompression and heat pumps are more common and depend on the process type. In the lower range, several technologies including variants of Rankine cycle (ORC, Kalina, or supercritical operation), several heat pump types and ejectors may harness further waste heat. Heat exchangers cover all the application ranges.

## 4.1. High and Medium Temperature Range

High temperature processes in industry are often quite inefficient. They generate a significant amount of heat by fossil fuel combustion, chemical reaction or other mechanisms.

Compared with the low-temperature waste heat, waste heat at high temperature is recovered more easily due to its high energy level, which could meet the user demand with different temperature requirements. The high-temperature waste heat can also be used for power generation with relatively mature technologies such as steam turbines or organic Rankine cycle (Z. Y. Xu, Wang, and Yang 2019). Therefore, in industries where high and medium temperature waste heat sources are present, it is relatively easy to capture substantial amounts of this heat and find matching sinks for its effective use. The higher is the temperature and therefore the quality of the heat stream, the easier to recover a larger amount of otherwise wasted heat. However, there may be some limitations related to stream types, elevated temperatures and fluid cleanliness, which may require specific technologies, some of which are still emerging so that further research is ongoing for improved capture technologies. A wide range of WHR technologies can be applied to capture this energy in its different forms and conditions. Technology consist mainly of recovery systems comprising recuperators, regenerators, boilers and several others devices. For further details, a comprehensive review by Jouhara et al. (2018) covers common sectors of industry for waste heat and its recovery, the predominant technologies, the use of units such as heat pumps and systems handling of waste heat. In other instances of extreme conditions, current recovery systems may not be efficient enough and new solutions or devices need to be developed. The paper by Zhang (2019a) illustrates the case of the waste heat generated from a coke-making process. This resource is abundant and its recovery may be usefully employed in many applications but sensible heat from coke oven gas (COG) cannot be recovered in efficient manner by empirical approaches, which are to re-combust residual heat, purify hydrogen and synthesize methane for energy generation. The authors proposed to alleviate this problem by the introduction of the Ascension-pipe heat exchanger (APHE).

A pilot study of the APHE was implemented. The average effectiveness was found to be up to 69%, demonstrating that the APHE was a reliable choice to recover sensible heat from COG and use it for steam generation. Further information on high temperature waste heat, its means of recovery and main barriers in harsh environments can be found in a recent article by Vance et al. (2019). This gives a detailed account on past and current technology developments on WHR and paths towards advanced materials, design research and development in this domain.

## 4.2. Low-Grade Energy

An overview of the available literature on waste heat classification in terms of temperature levels broadly indicates that 50% correspond to low-grade (low temperature) waste heat, more or less ranging from 30°C to 250°C, as already mentioned in the above developments.

Recovery of low-grade thermal energy, even though viewed as a great opportunity to enhance efficiencies at various stages of processes and control their carbon footprint, presents special challenges. For instance, when the recovery temperature decreases and gets closer to the ambient, the exergy content reduces, heat transfer becomes less efficient and irreversible losses increase. In these conditions, conventional power generation systems are no longer operational. Moreover, matching difficulties may arise to find use for the recovered heat as suitable user demand may not always be available. Z. Y. Xu, Wang, and Yang (2019) summarized these issues in three main points: (1) lack of global optimization methodology which can integrate the heat exchange and energy conversion simultaneously; (2) distributed waste heat recovery needs high investment; (3) mismatches between the waste heat supply and demand in time, space and energy grade, limit the potential of waste heat recovery. The potential applications commonly contemplated for these conditions generally consist in upgrading the waste heat by means of a heat pump for use in domestic hot water, space or process heating (Ziviani, Beyene, and Venturini 2014). More information about barriers, concerns and options regarding low-grade waste heat recovery considerations may be found in Jouhara et al. (2018). These important elements substantially affect the adoption of low-grade waste heat applications. In addition, Varga and Palotai (2017) raised the fact that despite the improvement of methods for process/process heat exchange,which have been applied for a long time, the general practice for utilizing energy content of low streams temperature has not yet been sufficiently developed in terms of engineering design. In this respect, their paper was an illustrative example of an engineering study to partially substitute an air cooler in a crude oil refinery unit with ORC and Kalina systems to generate power with low-grade heat. The results indicated a potential gain in efficiency with either replacement technology of over 10%.

In their review of ORC system modeling, Ziviani, Beyene, and Venturini (2014) made a similar observation for this technology, in that most research on low-grade heat recovery focused mostly on mathematical modeling and feasibility studies but the little progress on the practical side leaves doubts about engineering and practical limitations of these concepts. Consequently, there is a lack of suitable cost-effective technology ready for mass production. Their paper focuses on ORC challenges, methods and best practices for better design solutions.

Renewable energy, such as solar, wind, biomass and shallow geothermal sources are also included in the low-grade energy classification and characterized by the supply of energy on an unlimited basis. A great deal of interest is being put nowadays on these sources as future alternatives to the conventional energy supply structure (Panwar, Kaushik, and Kothari 2011). Consistent efforts to exploit these sources in applications of interest such as desalination, cooling and heating or electrical and mechanical power are increasing. The work of Lund and Persson (2016) with regards to mapping potential sources for district heating in Danmark is a good example of low-grade heat recovery project in the framework of a 100% renewable supply by the year 2050. The authors analyzed eight heat source categories for a heat-pumping network (low-temperature industrial excess heat, supermarkets, waste water, drinking and usage water, ground water, rivers, lakes and sea water). A conclusion was reached in that potential heat sources are present near almost all district heating areas and that sea water most likely will have to play a substantial role as a heat source in future energy systems.

In buildings, the use of WHR systems increases to transfer low temperature waste heat from laboratories, data centers or other sources to provide beneficial heating or cooling in other parts of the building. There are mainly two media in building systems to displace waste heat: air and water. Water WHR systems reuse the heat produced by cooling/heating operations that is usually rejected for example by a cooling tower or a condenser. Air heat-recovery systems are often used to preheat and/or precool ventilation air to reduce the load in residential building. A typical heat recovery system is composed of a heat exchanger, a fresh inlet, a contaminated air exhaust, and a fan (Figure 42) (Q. Xu, Riffat, and Zhang 2019; Mardiana-Idayu and Riffat 2012). Such a system can recover about 60-95% of waste energy/heat (Q. Xu, Riffat, and Zhang 2019).

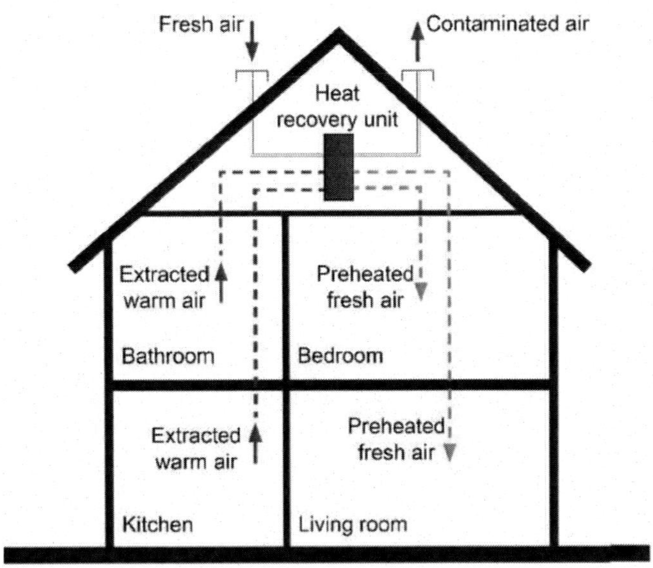


Figure 42. Typical heat recovery system for a residential building (Q. Xu, Riffat, and Zhang 2019).

## 5. Technologies for Heat Recovery

Ideally, for maximum process efficiency the best approach would be to minimize waste heat generation and maximize its recovery. As aoutlined in the previous paragraphs, waste heat was shown to be abundantly available and generated in industrial processes in a wide range of temperatures and cleanliness or naturally available as a renewable, low grade energy.

Typically high temperature waste heat results from direct combustion processes, medium temperatures from exhaust combustion units and low temperatures from parts, products and process equipment units, in addition to the renewable heat sources. Correspondingly, a large variety of conventional technologies is available for its maximized recovery or technologies at research and development stages in order to best adapt them to temperatures and conditions of emitting sources. A detailed discussion of the waste heat recovery technologies is beyond the scope of this work and only highlights on the main conventional and emerging heat recovery devices and systems will be explored.

In this way, there are technologies and techniques typically used in the higher temperature range, such as heat pipes, regenerators of waste heat from coke ovens, oven chambers, blast furnaces, electric power generation and more. The medium range of temperatures generally correspond to systems such as economizers, rotary regenerators, recompression systems and heat pumps, ORC and Kalina cycles. In the low temperature range, mechanical and thermal heat pumps, ORC and Kalina cycles or thermally activated refrigerators are common as well. Some technologies may be applicable in more than one temperature level such as heat exchangers, heat pumps or ORC's. For more information, the review papers of Vance et al. (2019) and Jouhara et al. (2018) are especially dedicated to the issues of waste heat and give an extensive coverage of the related recovery technologies.

Ejectors on the other hand, have been in application for many years (ASHRAE 1965), but they are mainly restricted to industrial applications using steam as the working fluid. They are commercially available in different sizes and geometries to match different power requirements.

In this chapter, a typical steam ejector application is described for the sake of illustration but the important potential of this technology with working fluids other than water will be the subject of a complementary development in a separate chapter.

## 5.1. Conventional Equipment

### *5.1.1. Heat Exchangers*

A large number of engineering applications involve an exchange of heat between two fluid streams at different temperatures, separated by a solid wall to avoid cross-contamination. Such a process is performed by a heat exchanger which is a passive device, allowing the effective transfer of quantities of heat without mixing. There are many designs of heat exchangers performing different functions in industrial, commercial and domestic areas. In industrial waste heat recovery processes, flue gases are often the heat source of high temperature, which is typically used to heat other gases or preheat liquids, including process water. Over a wide range of temperatures and applications, heat exchangers serve as steam generators, condensers, evaporators, heaters and coolers. They play a major role in industrial processes, power generation systems, heat pumps, refrigerators and so on. There are many heat exchanger designs corresponding to a considerable variety of application conditions but compact heat exchanger designs are nowadays becoming predominant, due to their high efficiency (up to 90%), low volumes in the range of 5% to 50% of their traditional shell and tube equivalents of the same duty. They have a high area density (ratio of heat transfer surface to volume), ranging from 300 $m^2/m^3$ for liquids or two-phase streams to 700 $m^2/m^3$ for gases. Moreover they generally have moderate costs overall and can work in harsh or special conditions. There are many types of compact heat exchangers, starting with the plate and frame, once used in food and dairy industry but now more widespread with brazed and welded plate versions available for a wide range of operational environments. Other heat exchanger designs such as spiral heat exchangers, compact shell-and-tube heat exchangers, printed circuit heat exchangers and many more are now available for a wide range of uses. For interested readers, a wealth of details on the subject may be found in Reay (1999); Cuce and Riffat (2015); Al-Rabghi et al. (1993) and Hesselgreaves (2001).

In general and more especially for cases where a change of phase is present (i.e., generators, evaporators and condensers), it is recommended to operate at the smallest temperature possible (pinch). Even though this parameter varies with heat exchanger design and the application, it plays an important role on the size and operation of the device (size, pressure losses and costs). Typical values reported in the literature are 4-8°C in evaporators and 2-6°C in condensers for the liquid heat sink and heat source, respectively. If dealing with gases, then these differences are higher. Design pressure drops are typically 2°C and 1-2°C respectively for evaporators and condensers (IEA 1995).

In certain conditions, new compact and efficient plastic heat exchangers offer an alternative to metal heat exchangers often subject to corrosion, oxidation and microbiological attack. They thus experience slow degradation and loss in heat transfer capacity when used in some chemical process. In livestock barns, they save energy and money while reducing greenhouse gases.

### *5.1.2. Heat Pipes*

Heat pipes and thermosiphons are included in the category of passive heat transfer devices as well, since they fulfill similar duties as those of heat exchangers.

A thermosiphon in this respect is a natural two-phase convection device (Figure 43) allowing the transfer of heat from a heat source (evaporator at bottom) to a heat sink (condenser at top) separated by an adiabatic zone. A typical unit is a single vertical tube or container (Figure 43a), sealed at both ends, with a small charge of two-phase working fluid, but configurations are adapted to suit intended applications. The container is mostly filled by the vapor phase of the working fluid, while the liquid phase makes up the rest of the volume. A closed-loop thermosiphon works in the same manner as the single vertical thermosiphon, except that two tubes, the riser and the downcomer connect the evaporator and the condenser instead (Figure 43b). In this case, the descending liquid and the ascending vapor do not interact.

A heat pipe differs from the thermosiphon only by an internal wick against the wall for moving the condensate back to the evaporator by capillary action. This gives more flexibility of orientation to the device, which does not need to be vertical. In the case of a thermosiphon, condensate return occurs solely by gravity. When the liquid absorbs heat in the evaporator, it flows upwards as vapor to the condenser where heat it transforms back into liquid after releasing its latent heat content. These devices constructed from suitable metals and working fluids, depending on the application, are characterized by effective thermal conductivities ranging from 5000 to 200 000 W/m K.

A relatively new tendency in thermosiphon technology is the introduction of nanotechnologies (Vasiliev et al. 2013), opening up new niches, particularly for low-grade temperature WHR applications, in acidic and saline process environments.

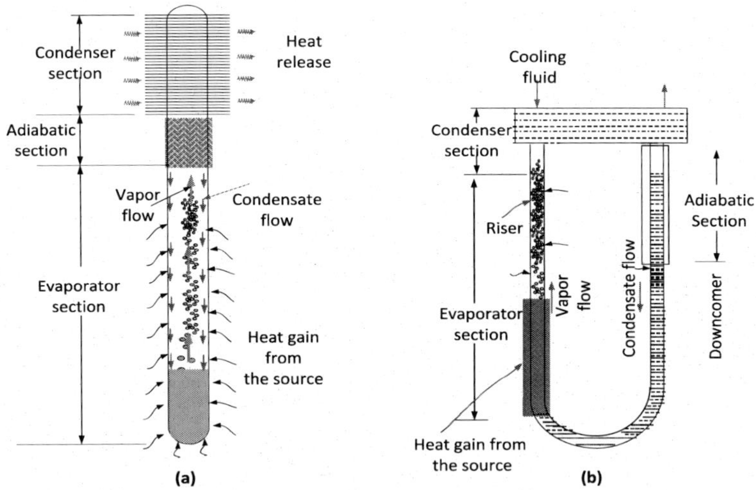


Figure 43. Schematic views of (a) single tube thermosiphon, (b) closed loop thermosiphon.

Further details about thermosiphons and heat pipes, their particularities and applications can be found in many references (H. Zhang et al. 2018; Srimuang and Amatachaya 2012; Badache et al. 2019), to cite only a few.

### *5.1.3. Mechanical Heat Upgrading*

#### **5.1.3.1. Sub-Critical Operation**

Depending on the conditions, mechanical heat upgrading may be achieved by means of industrial heat pumps in a closed mechanical compression cycle or by mechanical vapor recompression (MVR) which in a semi-open cycle working with steam. Both technologies rely on a compressor generally driven by electricity.

Industrial heat pumps offer a high potential to recover waste heat, improve energy efficiency, reduce costs, and increase production capacity of systems and processes while cutting GHGs by minimizing the fuel consumption. They are used to produce low-pressure steam from waste heat extraction or upgrade low-pressure steam to high-pressure steam, which can reduce the process energy consumption of boilers, furnaces and other equipment. The simplest configuration of a closed cycle mechanical heat pump as shown in Figure 44, circulates a working fluid (refrigerant) through an evaporator, extracting heat from the source by evaporation. A compressor elevates the vapor pressure and temperature, which is then transferred to a heat sink where it is condensed, delivering the upgraded heat at this point for applications such as space or water heating. The condensate eventually undergoes re-expansion to the evaporator conditions to complete the cycle.

Actual industrial heat pumps rely on upgraded versions of this simple cyle by the inclusion of internal components such as sub coolers, economizers, intercoolers or multistage and cascade configurations (Chua, Chou, and Yang 2010). A review update by Arpagaus et al. (2016) revisits many cycle configurations and highlights the major advantages and challenges of the technology. Various industrial, commercial and domestic applications as well as ejector-integrated cycles were discussed.

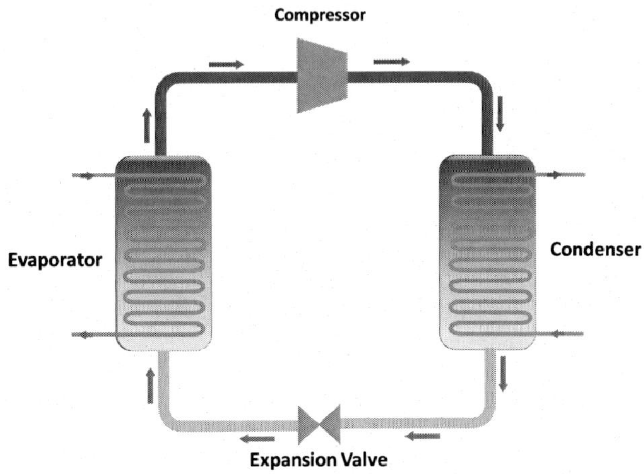


Figure 44. Mechanical single stage heat pump.

Maximum delivery temperatures by closed cycle heat pumps are limited primarily by the refrigerants and compressors and are expected to reach up to 130°C. The current state-of-the-art and the research activities on high temperature heat pumps (HTHP) were reviewed by Arpagaus et al. (2018) for heat sink temperatures in the range of 90°C to 160°C, hinting that these systems may soon become available. According to Nellissen and Wolf (2015), HTHP can work in the temperature range of 80°C to 150°C where there is a significant amount of heat demand in the chemical, paper, food, wood and other industrial sectors in addition to pasteurization, drying, distillation, sterilization, pressurized hot water production and other industrial processes. A summary development of a hydrocarbon HTHP for heat delivery up to 115°C from the theoretical conception to the experimental results is provided by Bamigbetan et al. (2017). However, commercially available heat pumps can generally supply heat only up to approximately 100°C.

MVR is the process of increasing the pressure and temperature of steam in order to reuse it. There are two MVR cases generally encountered and shown in Figure 45. In the first configuration called direct system, the steam from a low heat source is compressed directly, then condensed for use at higher pressure. This method is appropriate when contamination is not a concern. In the second case the configuration is reversed. The low heat source is used to vaporise a liquid before compression and use it as higher grade steam. This approach prevents mixing contamination. For low temperature lifts, MVR can be operated at low condensing temperatures in the range of 60°C to 80°C (IEA 1995). These systems have the advantage of efficiency but handling low-grade steam implies large component sizes and possibly high costs (O'Rielly and Jeswiet 2015).

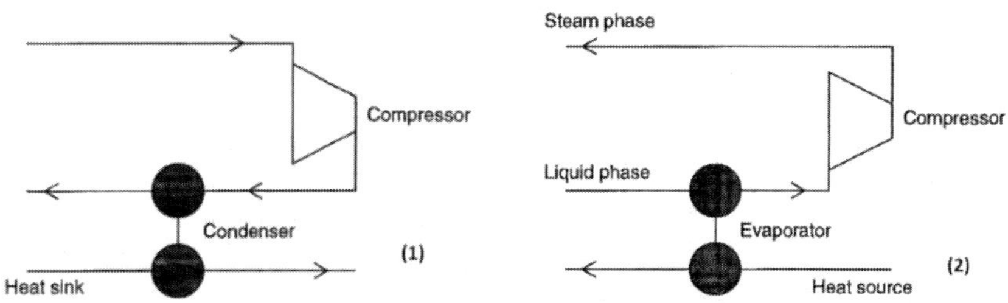


Figure 45. The two configurations of the mechanical vapor recompression, (1) direct system (2) reversed system (IEA 1995).

### 5.1.3.2. Transcritical Operation

A transcritical thermodynamic cycle operates with a working fluid conditions crossing the saturation line and varying from subcritical to supercritical conditions. Thus, a transcritical heat pump is built on this principle and carbon dioxide ($CO_2$) is the most commonly used working fluid in transcritical conditions (Figure 46). Refrigeration based on $CO_2$ is not a new idea, it was used before the advent of hydrocarbon refrigerants, its revival in the nineties was prompted by environmental concerns. The trans-critical cycle in its simplest form with no internal heat recovery is less efficient than sub critical cycles using ordinary working fluids. However, the thermophysical properties of $CO_2$ and its low critical point (corresponding to about 31°C) may be used to advantage on the heat sink side. This is because the heat sink is usually above the critical point (up to 40°C) where the pressure and the temperature vary independently and for a constant pressure above critical, $CO_2$ temperature varies with a substantially high lift (about 40°C is typical). This is particularly favorable for water heating applications, due to good matching between $CO_2$ and water temperatures in the gas cooler (Van de Bor, Infante Ferreira, and Kiss 2015). Heat extraction in the evaporator needs to occur at temperatures sufficiently low and away from the critical point for the heat pump to be efficient. This is particularly interesting for recovering ground source energy for space and water heating as demonstrated by Badache et al. (2018) for example. In addition, internal heat recovery will minimize expansion losses, particularly important for $CO_2$, and enhance the overall cycle efficiency above that of a conventional heat pump. A trans-critical cycle is characterized by high pressures on the gas cooler side but this disadvantage may be counterbalanced by small volumes and size components when it comes to safety and costs considerations.

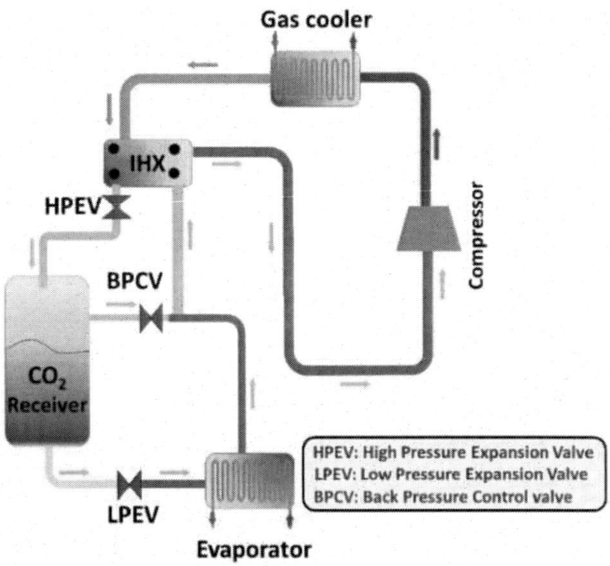


Figure 46. Transcritical $CO_2$ single stage heat pump with an internal heat exchanger.

## *5.1.4. Cycles for Thermal Heat Upgrading*

### 5.1.4.1. Steam Ejectors

In addition to mechanical recompression discussed in the previous section, thermal vapor recompression by mean of ejectors (Figure 47) has been in practice in industries where distillation or evaporation with moderate lifts (10°C to 20°C) were needed. This heat pump is driven by heat and no electricity is required, but needs a relatively high steam pressure to be activated (7 bar to 15 bar). It is of simple design, has no moving parts and its main advantages are low capital and operating costs (IEA 1995). Steam jet refrigeration was also common for cooling and air conditioning for residential commercial and institutional buildings and in industry until the advent of HFC refrigerants, which favored the spread of mechanical compression refrigeration and the decline of jet refrigeration, except in restricted niches of industrial applications.

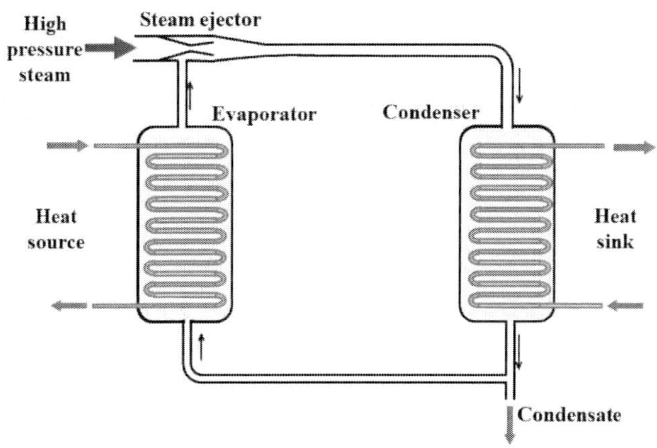


Figure 47. Thermal vapor recompression by mean of steam ejector.

### 5.1.4.2. Absorption Cycle Types

Similar to steam jet systems, absorption cooling and heating technology relies on heat for its activation but ordinarily employs a binary mixture as a heat carrier, instead of using a single component working fluid. This mixture consists of a volatile fluid or refrigerant and a nonvolatile component, which is an absorbent such that at a given pressure, the mixture boils at a higher temperature than does a pure fluid. Two binary fluid solutions have so far, been extensively used in absorption heat pump (AHP) technology. The fluid $LiBr/H_2O$ is more frequently used in industrial applications and moderate cooling (temperature range above 0°C). The $NH_3/H_2O$ pair also finds use in refrigeration. Different working pairs, including the water based working pairs, ammonia based working pairs, alcohol based working pairs, and halogenated hydrocarbon-based working pairs are classified by Z. Xu and Wang (2017), based on the refrigerant.

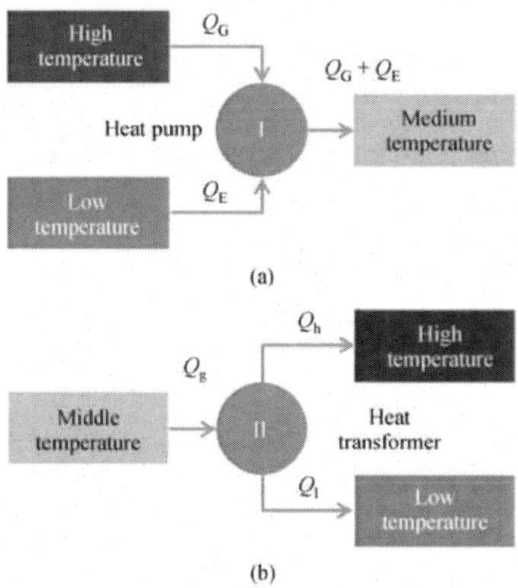


Figure 48. Absorption heat pump configurations, (a) Type 1 and (b) Type 2 (Z. Xu and Wang 2017).

The waste heat source for an absorption machine can be low-pressure steam, a hot gaseous or liquid stream. This is favorable since a substantial percentage of industrial waste heat is in the forms of hot water, steam, flue gas, etc. This technology is able to capture and deliver a useful heat output at different temperatures. In the review paper of Z. Xu and Wang (2017), advanced absorption cycle configurations including heat amplification, heat transformation for temperature upgrading and other multi-effect AHP cycles are described. In addition, several AHP and heat transformer applications in different industries are introduced. Conventionally the absorption pump in its simplest configuration is nearly composed of one set of components but depending on the heat flow directions between high and low pressure, two heat pump configurations may be distinguished (Figure 48). The first configuration depicted in (Figure 48a) is the Type 1 AHP. In this case, heat is lifted from low to medium temperature by supplying a certain amount of heat at high temperature.

Figure 48b is the heat transformer configuration or Type 2 AHP. Here, pressure levels are reversed and, the heat supply occurs at medium temperature. Part of this heat upgrades to a high temperature, while the remaining part is delivered at low temperature. The use of either cycle depends of the requirements of the process and the availability of waste heat at different levels.

### *5.1.5. Thermodynamic Power Cycles*

#### **5.1.5.1. Rankine Cycle**

A steam power cycle for mechanical power generation is perhaps among the most needed and useful ways to recover high temperature waste heat. It relies on the Rankine cycle, introduced here in its simplest form (Figure 49). Waste heat is used in a boiler heat exchanger to vaporize water to superheated steam, which is then sent to a high-pressure steam turbine where it is expanded to a lower pressure for the delivery of mechanical power. After leaving the turbine, the steam condenses before being pumped back to the generator. The power delivered by the turbine may then drive an alternator for electricity production. A first remark is that the efficiency of this simple cycle depends on average temperatures of the heat supply and the heat release, so that any measure to increase the input temperature or to decrease the output temperature will have a positive effect on the cycle efficiency. As a result, several improvement measures of the cycle were developed, the conventional ones are available in most applied thermodynamics handbooks like Borgnakke and Sonntag (2012). Configurations based on this cycle use the steam generated at high temperature by waste heat boilers and heat recovery steam generators for power/electricity generation (Jouhara et al. 2018).

Power plants operating on supercritical Rankine cycle (pressure exceeding the critical pressure of water, 221 bar and temperature higher than 600°C) offer a better thermal efficiency than those running on a subcritical cycle, but high pressure and temperature stresses impose special metallurgical considerations (Tumanovskii et al. 2017).

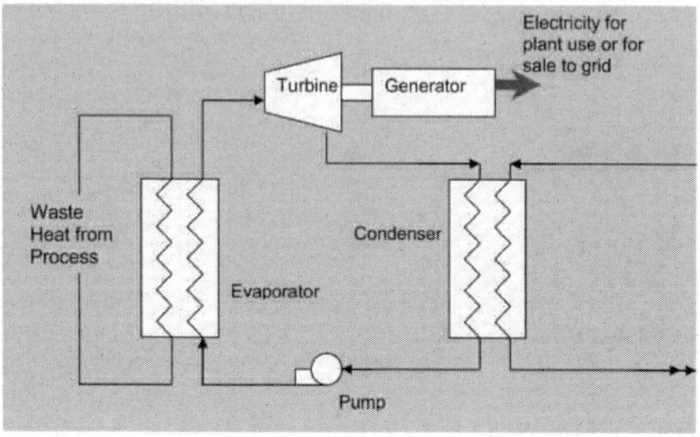


Figure 49. Waste heat recovery with Rankine cycle (USDOE 2008).

However, a Rankine cycle using steam is generally more suitable for high temperature applications and large centralized systems, a condition not necessarily always found on waste heat sites where most heat is medium to low-grade for which this cycle cannot provide good performance.

At medium to low temperature, suitable low boiling refrigerants may, on the other hand, be considered for the supercritical Rankine cycle (Shengjun, Huaixin, and Tao 2011; Lecompte et al. 2015).

Furthermore, the interest of this cycle structure has close similarities with organic Rankine cycle and other derivatives, which are introduced next paragraphs.

Supercritical and subcritical Rankine cycles share the same general lay out but they differ in the way the heat process is managed (Figure 50b-c). Instead of vaporizing the fluid to a two-phase, subcritical condition, the working fluid is compressed directly to a supercritical pressure. It is then heated to a supercritical temperature, thereby bypassing the isothermal phase change process. As a result, a temperature glide of the refrigerant favors a better matching between the heat source and the working fluid profiles, with expected efficiency improvements.

Using supercritical $CO_2$ in a Rankine cycle (Figure 51) may be a promising pathway. Sarkar (2015) discussed the potential of this cycle, in comparison to conventional steam and organic Rankine cycles. In addition to superior performance, this system offers advantages in terms of compactness as well as additional environmental benefits.

### 5.1.5.2. Organic Rankine Cycle (ORC)

As stated in the discussion of Rankine cycles, medium to low-grade heat (80°C to 400°C approximately) cannot be efficiently exploited by the technology. The ORC on the other hand, can be used to exploit heat sources in this range for power generation. Industrial waste heat, renewables such as solar, geothermal or biomass energies, all fall in this range of temperature. A schematic representation of a simple ORC cycle is shown in Figure 50. It is analogous to a classical Rankine cycle for steam but uses, low boiling point organic fluids for its operation. The main components of this cycle are the evaporator, the turbine, the condenser and a pump. To these components is generally added an internal recuperator, before the condenser and the evaporator inlets, to improve the overall efficiency. The preheated working fluid enters the evaporator and leaves as high-pressure vapor to expand in a turbine and generate power. The expanded vapor leaves the turbine at low pressure and temperature and cools down further by releasing more heat in the recuperator before the condenser. The liquid refrigerant leaving the condenser is pumped back through the recuperator where it is preheated by internal heat recovery before returning to the evaporator, finally completing the cycle. The ORC cycle architecture is diversified and the main features of several configurations are described in detail by Lecompte et al. (2015) and Mahmoudi, Fazli, and Morad (2018).

Similarly to the Rankine cycle for steam, the temperature is highest in the evaporator and lowest in the condenser. However, organic working fluids, due to their vapor pressure properties can work at relatively low temperatures and reasonable pressures. In addition, generally due the reduced specific volume of the organic working fluids, ORC turbines have small sizes, provided that the working fluid is correctly chosen. The selection of the working fluids for the different heat recovery technologies are discussed in some detail in a separate paragraph. It is worth to recall here that the critical temperature, stability, specific volume and a dry or isentropic saturation characteristic curve are among the important criteria to keep in mind in most cases (Tchanche et al. 2011; Aidoun et al. 2019; Jouhara et al. 2018).

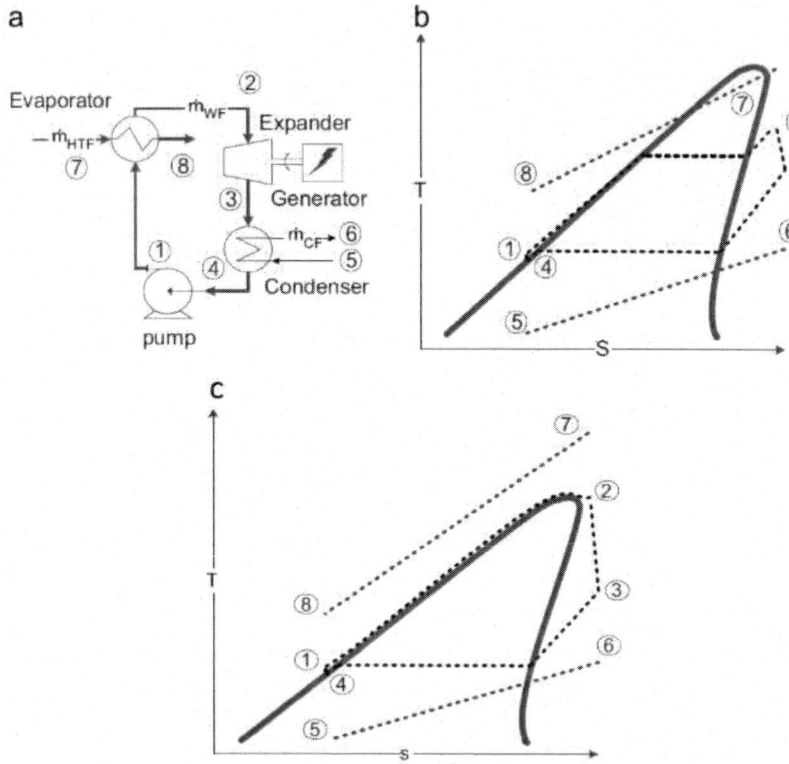


Figure 50. (a) ORC cycle layout, (b) subcritical T-s diagram, (c) supercritical T-s diagram (Lecompte et al. 2015).

This technology offers a considerable potential to harness medium to low-grade waste heat, which constitutes the highest portion of the total amount released, and convert it to power. In this range of temperatures, ORC conversion efficiency is much lower than conventional Rankine cycle system technology, which is around 30%. Theoretical efficiency is estimated to be in the 6-16% range and experimental evaluations lie between 4 and 5.6% (Vélez et al. 2012; Varga and Palotai 2017). However, due to its wide-ranging potential, ORC can as well recover waste heat from other power cycles (turbines, engines, etc…) and even combine with other technologies such as chillers, heat pumps or heat recovery cascades. Moreover, unlike conventional power-cycles, which are large scale, this technology allows for decentralized and small-scale power generation.

Subsequent reviews by Vélez et al. (2012); Quoilin et al. (2013) Ziviani, Beyene, and Venturini (2014) are worth mentioning. The first two give a detailed account of ORC technology with an emphasis on applications, technical, economic and market aspects of the technology. The importance of the working fluid type selection is highlighted, recalling that no single fluid can fulfill all requirements over the wide range of conditions covered by ORCs. The authors covered a wide range of applications, currently limited to a power generation capacity of 0.2-2 MWe. Focus was put on technical advantages and limitations, complemented by a concise economic and market analysis, cost figures for different commercial ORC modules and manufacturers and the current trends in research and development for preliminary evaluations. Energy sources included in the analysis were solar for electricity and desalination, biomass in combustion, gasification or biogas forms, geothermal and process waste heat.

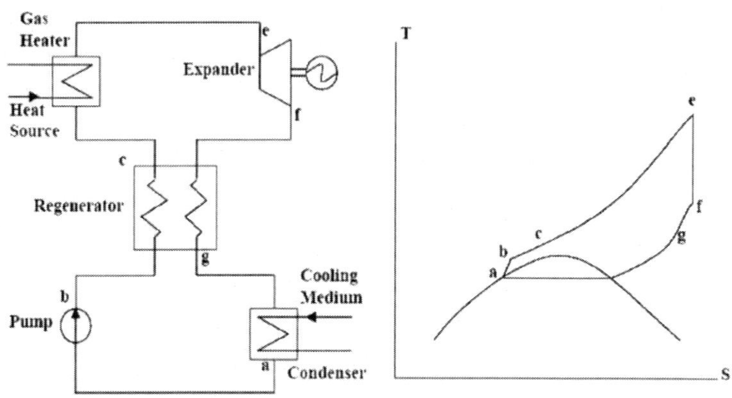


Figure 51. Supercritical $CO_2$ Rankine cycle with internal heat exchanger (Sarkar 2015).

A survey of common modeling approaches in the literature, under both steady and transient conditions is available in Ziviani, Beyene, and Venturini (2014). The information includes ORC system component modeling, challenges and best-practice scenarios for better design solutions. This information is particularly important at the early stage of projects to evaluate the potential of the ORC implementation.

A comprehensive estimate of ORC applications to energy intensive industries like steel, glass and oil in the EU is presented in Campana et al. (2013). Noting that, authors in their methodology, intended for both new plants and those already in operation showed that in the targeted sectors, about 20,000 GWh of thermal energy per year could be recovered, allowing to reduce GHG emisions by 7.6 Megaton of $CO_2$ equivalent.

Other modeling studies are rather prolific about ORC design and parametric optimization for different configurations and conditions. Maraver et al. (2014) investigated thermodynamic optimization with different average heat-source profiles in view of providing guidelines for a wide range of operating conditions of subcritical, trans-critical, regenerative and non-regenerative cycles.

A database of information on ORC technology is available in Landelle et al. (2017). This database is built on over 175 experimental studies in the literature with the following features: ORC components, main parameters performance criteria, and heat source type. The database enables the comparison of ORC cases based on the identification of similar installations, irrespective of whether data is collected from the same reference or not.

Special applications forming ORC hybrids with other technology types are found more frequently in the literature, generally as feasibility investigations. A typical example is the study of the ORC-Gas turbine hybrid (waste heat recovery from gas turbine) proposed for an off-shore oil and gas platform application by Pierobon et al. (2013). The authors compared ORC with two further technologies, i.e., air bottoming and steam Rankine cycles. Findings suggested that the ORC unit was best in terms of performance and economic revenue, ensuring a combined cycle thermal efficiency of 41.5% and a payback on investment of around 5 years.

Similar results of ORC application to internal combustion hybrids were compiled by Sprouse and Depcik (2013) who reported fuel consumption reduction of up to 10%. Along the same path, a review paper was recently published by Chintala, Kumar, and Pandey (2018) on waste heat recovery from a compression ignition (CI) engine by means of an ORC, indicating that harnessing heat at different temperatures (from exhaust and cooling fluid) was feasible. Thermal efficiency of a standalone engine-ORC system is estimated to be in the range about 10-25% and the overall efficiency of the combined system is significantly higher than the conventional system. They also pointed out that high capacity (power output) multi-cylinder CI engines, which deliver a high mass flow rate of exhaust gas at high temperatures, are the best suitable ones for integration of ORC systems.

Another recent survey by Mondejar et al. (2018) was devoted to the use of organic Rankine cycles for waste heat recovery on board of ships, providing an analysis of the potential and the challenges of using this technology in retrofitting current vessels. Application to the ship's waste heat recovery by the authors suggest potential fuel consumption reductions in the order of 10-15%.

#### 5.1.5.3. Kalina Cycle

A Kalina cycle is very similar in structure to an ORC cycle and has been described in many papers (Varga and Palotai 2017; Sprouse and Depcik 2013; Van de Bor, Infante Ferreira, and Kiss 2015; Jouhara et al. 2018) to mention only a few. This modified cycle version includes, in addition to conventional components found in ORC, an absorber and flash tank, as represented in Figure 52. Moreover, instead of a single component organic fluid generally used in an ORC, the Kalina cycle ordinarily uses a binary ammonia-water compound and less frequently a binary mixture of hydrocarbons depending on applications. This principle relies on the fact that when two fluids of different boiling points are mixed to form a compound, this later boils or condenses with a temperature glide for a given pressure, while the respective concentrations of the mixture components vary during the evaporation (or condensation) process. It becomes possible in this way to ensure a good operation flexibility in a range of temperatures by varying stream concentrations and thus maximizing heat extraction from the waste heat source. The ammonia-water mixture particular advantage in favor of its more common use is that its molecular weight is close to that of water so that a standard back-pressure turbine can be used (Jouhara et al.

2018). In this case for example, when an ammonia/water mixture is heated, ammonia vaporizes first since it is more volatile than water. Ammonia concentration in the liquid decreases and corresponding saturation temperature increases, resulting in a temperature glide that may be used to adjust fluid evaporation and heat source temperature matching for improved overall cycle efficiency.

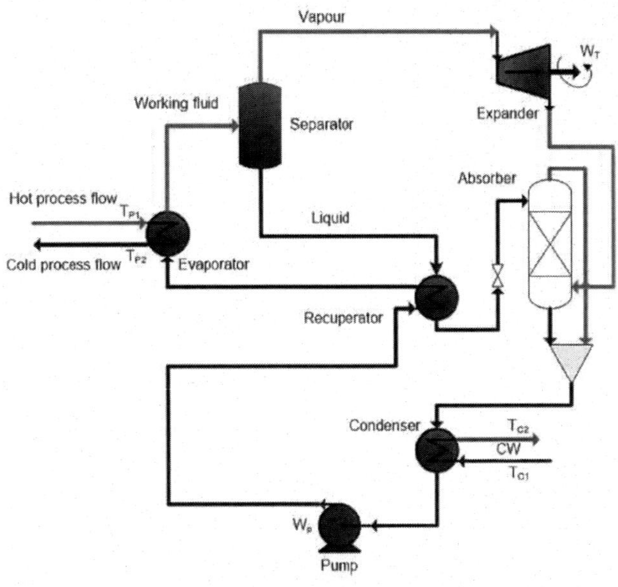


Figure 52. Kalina cycle (Varga and Palotai 2017).

Referring to Figure 52, in the operation of this cycle a cold liquid mixture with a high concentration of ammonia is pressurized by a feed pump is first preheated with the hot, weak ammonia solution from the absorber, prior to entering the evaporator. Heat is extracted from the source of waste heat and transferred to the rich liquid mixture from which part of the ammonia content is released, superheated, sent to a turbine and expanded while the poor, liquid solution is sent to the absorber after separation. The ammonia vapor leaves the turbine at lower pressure and temperature to be absorbed with the weak ammonia solution and cooled down before starting the new cycle. Because of the ability of maximizing waste heat extraction from the available source, the Kalina cycle is generally more efficient than an ORC in the high-medium range of temperatures. However, for low-medium temperature levels of the heat source, the gains in efficiency are sensibly similar to those of an optimized ORC with less complexity (Jouhara et al. 2018). Such a comparison was carried out by Varga and Palotai (2017), who performed their study on a crude oil refinery unit heat source with ORC and Kalina operated under the same main constraints but optimized for maximized efficiency, heat and cost recovery. It was found that the efficiency of the selected systems at the highest power generated was 10.0% for ORC with i-pentane as working fluid and 10.57% for Kalina cycle with ammonia-water mixture, within the design boundaries (corresponding to pressures of 29 bar and 7 bar respectively for Kalina and ORC cycles). The return on investment of both technologies were estimated to be approximately 5 years, showing that at this level of details, the study does not clearly differentiate between both alternatives.

#### 5.1.5.4. Trilateral Cycles

A trilateral cycle system lay out is similar to that of an ORC in terms of their main components. From the schematic representation shown in Figure 53, it may be seen that the main components of this cycle are the heater instead of an evaporator, the two-phase expander rather than a turbine, the condenser and a pump. In terms of process, the liquid is pumped from the condenser low pressure saturated condition to the high pressure of the heater where its temperature is elevated to the corresponding saturated liquid state. The saturated liquid is fed directly to the two-phase expander. In this way, isothermal boiling is avoided, thus allowing an optimized matching between the temperature profiles of the heat source and the working fluid for improved cycle efficiency. Studies indicate that even though this cycle has intrinsically a lower efficiency than the basic ORC, but its potential for better overall efficiency is higher, due to its ability to maximize heat recovery.

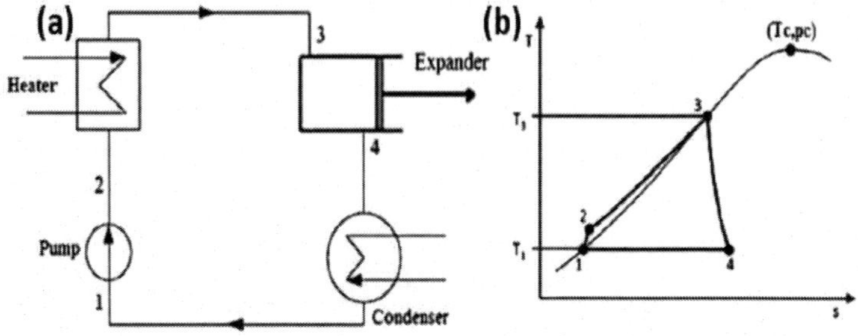


Figure 53. (a) Trilateral cycle layout and (b) T-s diagram (Ziviani, Beyene, and Venturini 2014).

Further studies suggest that this cycle could attain higher thermal efficiencies than the ORC (i.e., between 35% and 15% more for heat source temperatures of 115°C and 160°C, respectively) (Lecompte et al. 2015; Mondejar et al. 2018). However, this technology is still under development. More work remain to be done on component development, which are not readily available as efficient and reliable two-phase expanders.

## 5.2. Emerging Heat Recovery Devices and Systems

The search for technologies and methods to recover and make best use of waste heat remain intensive. Many efforts have focused on ways to convert waste heat directly into electricity, thus bypassing the transition through the conventional mechanical energy stage of the power cycles discussed in previous paragraphs. Technologies based on thermoelectric, piezoelectric, thermionic and thermo-photovoltaic generation of electricity are discussed by Jouhara et al. (2018) who reports that these technologies have still not penetrated the market but prototypes of a number of them have shown a promising potential.

A brief description of these devices is given here; for more information, the reader is redirected to the work of Jouhara et al. (2018).

The thermoelectric system is based on the Seebeck effect, which generates an electric current between two semiconductors, respectively subjected to a hot and a cold source of heat.

The piezoelectric power generation device converts low temperature heat energy directly to electricity. Made out of thin-film membranes, they transform ambient vibration such as oscillatory gas expansion into electricity.

In a thermionic device, a temperature difference between a cathode at high temperature (Emitter) and an anode at lower temperature (Collector) generates a flow of electrons between metal and metal oxide surfaces in an inter-electrode space under vacuum to generate electricity.

A thermo-photovoltaic generator directly converts radiant energy into electricity in a similar manner to a solar panels. They use an emitter, which, when heated by the heat source, emits an electromagnetic radiation. The PV cell converts the radiation to electrical energy and the spectral filter passes through only the wavelengths matching the PV cell.

These technologies despite their potential are generally characterized by low efficiency in addition reliability, stability and costs remain as the important issues, which have yet to be fully addressed.

## 6. Working Fluids

In the majority of heat recovery systems and devices, waste heat is captured by means of a heat transfer fluid circulating between the heat source (capture point) and the heat sink (delivery point). In heat pumps, ejectors and many cooling devices, the working fluid is generally referred to as the refrigerant, whereas in power generating systems like ORCs, Kalina and other cycles it is simply the thermal fluid. Independent of the terminology, the working fluid in any case is of crucial importance for the system and evolves with its developments. The main types of working fluid used in the literature (Bao and Zhao 2013; Saidur et al. 2011; Nasir and Kim 2016; Bamigbetan et al. 2017) are summarized in Figure 54. The following discussion deals with the role of the working fluids in power generation and heat pumps. More information on the role of the working fluids in the ejector technology is given in separate chapter. For further details about the other technologies, the reader is referred to technology specific reference (e.g., Badache et al. 2019 for the case of thermosiphon).

Figure 54. Main types of working fluid used in the literature.

## 6.1. Working Fluids and Environmental Context

Many of the conventional working fluids for heat recovery, refrigeration and heat pumps are no longer used or in the course of extinction, due to their environmental impacts. They are the synthetic fluids with substantial Ozone Depleting Potential (ODP) and/or Global Warming Potential (GWP). Even the more recent unsaturated hydrofluorocarbons (HFO) are not entirely safe as they may potentially release toxic substances and present still unknown effects on the environment (Bamigbetan et al. 2019). This has led researchers to look for new fluid alternatives, safe to the environment with equivalent or better performance features than currently available.

## 6.2. Fluids Properties and Requirements

The selection of a suitable fluid for an intended engineering application generally calls for some compromise between performance and environmental innocuousness. This stems from the fact that there is no single fluid suitable for all systems or conditions of operation. For this reason, available studies typically focus on a range of conditions for given technologies. Research evaluations of different fluids were conducted with this in mind, allowing a better understanding of the influence of the fluid nature on system performance. In this way, it was found for example that the critical conditions affected the coefficient of performance and the boiling points affected the volumetric flow and components sizes. Fluid safety (toxicity, flammability) and material compatibility, etc..., appear to be important evaluation criteria in addition to those of the environment. Irrespective of the technology and its application range working fluids must comply with the important performance requirement, in addition to specific criteria depending on the temperature range in which they are used and the technology to which they are intended.

Generally, the working fluid must be environmentally harmless in terms of ODP, GWP and Atmospheric Lifetime (ALT). It must have favorable thermopysical and thermodynamic properties (wet and dry types, latent heat, density, specific heat, normal boiling and critical points, viscosity, and thermal conductivity). For example, the specific heat, the viscosity and the thermal conductivity play an important role in cycle performance and heat transfer efficiency. Low boiling temperatures facilitate low temperature recovery, density influences components size and latent heat affects the specific circulation rate. A high critical point offers operation flexibility in subcritical cycles and a lower critical point facilitates supercritical cycle operation. In addition, the working fluid must be safe, stable and compatible with materials, readily available and low cost. Thermal and chemical stability are particularly important in the case of heat recovery applications like those encountered in ORC and Kalina cycles (Varga and Palotai 2017), thermosiphon, and industrial heat pumps.

Given the environmental and economic contexts, future developments on working fluids for efficient heat recovery cycles and systems will increasingly focus on not only natural refrigerants (hydrocarbons, carbon dioxide, water and ammonia) but also hydrofluoroolefins (HFOs) and hydrochlorofluoroolefins (HCFOs) which are becoming available. Research on the current generation of HFCs continues for performance at high temperatures, despite their negative impact on the environment. They are still common in refrigeration and heat pump applications but new compounds such as HFOs and HCFOs are becoming more available. Natural refrigerants including hydrocarbons (HCs) and their blending to form zeotropic mixtures with adequate temperature glides, will contribute to performance enhancement while satisfying environmental requirements.

In this paragraph, it is not meant to give a detailed account of working fluid requirements and conditions for every technology given their differences, objectives and ranges of application. Therefore, only highlights on common and important features are provided with a few examples related to each technology of interest. For further details, the reader is referred to technology specific developments of Huijuan Chen, Goswami, and Stefanakos (2010); Chua, Chou, and Yang (2010); Bamigbetan et al. (2017); Bamigbetan et al. (2019); Arpagaus et al. (2016); Arpagaus et al. (2018).

## 6.3. Working Fluids in Power Cycles

Given the importance of the working fluid on ORC operation and performance, some authors focused on working fluid characteristics and selection criteria such as Abadi and Kim (2017); Bao and Zhao (2013); Huijuan Chen, Goswami, and Stefanakos (2010).

For example, Huijuan Chen, Goswami, and Stefanakos (2010) proposed a detailed review of 35 working fluids for ORC applications, based on a set of selection criteria including thermophysical properties, stability, environmental impacts, safety and compatibility, availability and cost which they considered in the framework of both the ORC and the supercritical Rankine cycles. The main fluid type influences in terms of latent heat, density and superheating effectiveness etc..., were thoroughly discussed. For example, working fluids with high density and high latent heat were shown to provide high unit turbine work output, isentropic and dry fluids are preferred in organic Rankine cycles, but superheating is necessary for wet fluids. Fluids with low critical temperatures and pressures are potential candidates for supercritical Rankine cycle, which offers a better thermal match than ORC. However, this generally comes at higher pressure working conditions with potential safety concerns.

Along similar lines Shengjun, Huaixin, and Tao (2011) compared the thermodynamic and economic performance of an ORC and a transcritical Rankine power cycle in the conditions of a low-temperature geothermal power plant (80°C to 100°C). The theoretical investigation relied on an optimization procedure with thermal efficiency, exergy efficiency, recovery efficiency, heat exchanger area per unit power output (APR) and the leveled energy cost (LEC) as indicators. The analysis was performed with 16 working fluids. The main findings of this study suggest that the choice of the working fluid substantially influences the value of the optimized operation parameters, which are not all the same for different indicators. Analysis with refrigerants R123 and R125 indicates that R123 in subcritical ORC system yields the highest thermal efficiency (11.1%) and exergy efficiency (54.1%). Although the thermal efficiency and exergy efficiency of R125 in transcritical cycle is 46.4% and 20% lower than that of R123 in subcritical ORC, it provides 20.7% larger recovery efficiency. As a result, R125 in a transcritical power cycle turns out to be more effective in an ORC powered by low-temperature geothermal heat.

In the ORC context, the main guidelines for screening working fluids approximately coincide on many aspects with those of other waste heat recovery technologies. Papadopoulos, Stijepovic, and Linke (2010) proposed a design and selection method of working fluids based on computer aided molecular design (CAMD) and process optimization techniques. Optimum designs are then proposed for optimum ORC performance with respect to important economic, operating, safety and environmental indicators. In a comprehensive review recently conducted by Fakeye and Oyedepo (2018) for ORC technology, the authors established thirteen basic criteria, including:

- Thermodynamic performance.
- Positive or isentropic saturation vapor curve.
- High vapor density.
- Low viscosity.
- High conductivity.
- Acceptable evaporating pressure.
- Positive condensing gauge pressure.
- High stability temperature.
- Melting point.
- High safety level.

The last three requirements need to be given due consideration. A recent review, proposed by Dai, Shi, and Qian (2019) about working fluid thermal stability, includes theoretical and experimental considerations of the method and data for fluid decomposition with the influence on ORC systems. Given the generally high temperature working conditions in ORCs, stability considerations are therefore of utmost importance. The fluid must not decompose under the effect of temperature, which not only would alter its thermophysical properties but also release toxic elements and contaminate the system with residues.

In addition, weighting factors for each requirement are applied in the fluid selection according to the operating conditions (e.g., temperature range) and the cycle type, thereby affecting the efficiency of the system, the lifetime of important components (such as the turbine or the expander) as well as overall costs.

However, no matter how accurate the design is, there is no single working fluid that may satisfy all the required conditions in which any technology is used. This fact is illustrated in the work of Carcasci, Ferraro, and Miliotti (2014) who compared four different working fluids (toluene, benzene, cyclopentane and cyclohexane) in order to recover waste heat recovery from gas turbines by means of an organic Rankine cycle optimized by varying the main pressure of the fluid at different temperatures of the oil circuit. Moreover, the possible use of a superheater is investigated for each fluid in order to increase electrical power. It was found that fluid performance depended on the oil working temperature: cyclohexane is the best fluid for low oil temperature, benzene is the best choice for medium oil temperature, and toluene should be used for high oil temperature.

The effect of fluid type on ORC to convert waste heat energy to power was also investigated by Escalante, de Carvalho Junior, and Balestieri (2017) in the range of normal boiling points of -33.35 to 79.85°C. Five fluids were compared based on first and second laws of thermodynamics: two isentropic fluids (R11 and R12), two dry fluids (benzene and R113) and a wet fluid (R717). High system thermal efficiency was generally obtained for isentropic and dry fluids. It was also shown that thermal efficiency was generally higher for the fluid with the highest boiling point. Ammonia (wet fluid) showed the poorest thermal efficiency.

Very recently, Eyerer et al. (2019) investigated the applicability of new HCFOs as drop-in replacements for R245fa in an existing ORC systems. Preliminary evaluations in terms of performance and material compatibility allowed for selection of R1233zd(E) and R1224yd(Z). The three fluids were then experimentally compared in a range of operational conditions. R245fa and R1224yd(Z) were found to be more compatible with existing material of ORC than the R1233zd(Z). A performance comparison indicated that R245fa power output was superior by 9% and 12% to R1233zd(E) and R1224yd(Z), respectively, while R1233zd(E) thermal efficiency was 2% higher than that of R245fa and R1224yd(Z), respectively.

In the Kalina cycle which shares many common features with ORC, ammonia-water mixtures used as working fluid provide a better temperature match to the waste source than would be a pure fluid evaporating at constant temperature (case of ORC), because of the difference of vaporization temperature between water and ammonia (Varga and Palotai 2017). Ammonia-water mixtures and lithium bromide water solutions are used as well heat pumps and refrigerators.

## 6.4. Case of Industrial Heat Pumps

A mechanical heat pump system is another essential element among the conventional technologies offering economical alternatives for heat recovery from various sources. These devices have been using conventional refrigerants for a long time. Improving the performance, reliability and operational footprint on the environment has always been among the main concerns and also applies to refrigerant development as well. In the specific context of industrial high temperature heat pumps, the refrigerant selection is again a key factor for efficiency, performance, and safe operation. Arpagaus et al. (2018) list of requirements in the refrigerant selection for industrial heat pumps include:

- Thermal suitability (i.e., high critical temperature and low pressure for subcritical operation, low-pressure ratio and positive pressure at standstill).
- Environmental compatibility (i.e., low GWP and no ODP).
- Safety (i.e., non-toxicity, no or low flammability).
- Efficiency (i.e., high COP, minimal superheat, high volumetric heating capacity).
- Availability and low cost.
- Other factors (i.e., solubility in oil, oil mixture thermal stability, material compatibility, etc…).

A great deal of research with a series of refrigerants such as R134a, R410a, R407a, R433a, mixtures R32/R134a and R170/R290 was performed to substitute the abandoned CFC's and HCFC's (Chua, Chou, and Yang 2010). More recent work focusing on heat pump technologies with natural working fluids such as ammonia for high temperatures (> 90°C) are in different phases of technological readiness (Bamigbetan et al. 2019).

The compressor in high temperature heat pumps is the main limiting factor in terms of durability, due to high discharge pressures and temperatures. An experimental heat pump prototype of 20 kW tested by Bamigbetan et al. (2019) with R600 and a compressor designed to provide suction and discharge temperatures of up to 80 and 140°C, respectively. The experimental results indicated that the prototype could upgrade waste heat from 50 to 115°C. The machine demonstrated good operating stability and a promising potential for even higher delivery temperatures. Halogenated hydrocarbons are progressively being replaced by pure hydrocarbons (Varga and Palotai 2017). However, research on the current generation of HFCs continues for performance at high temperatures, despite their negative impact on the environment. They are still common in refrigeration and heat pump applications but new compounds such as HFOs, HCFOs are progressively made more available. Natural refrigerants including hydrocarbons (HCs) and their blending to form zeotropic mixtures with adequate temperature glides will contribute to performance enhancement while satisfying environmental requirements. Arpagaus et al. (2018) reported a variety of refrigerants currently used in heat pumps (e.g., R245fa, R717, R744, R134a or R1234ze(E), others continue to be studied in several research projects in view of achieving higher COPs at higher temperature heat sinks (R1336mzz(Z), R718, R245fa, R1234ze(Z), R600, and R601, R1336mzz(Z), Novec 649). New binary fluids like BY-5 are being considered for high sink temperatures. Experiments by (Y. Zhang et al. 2017) with BY-5 in the range of 70-80°C evaporation and 110-130°C condensation are promising.

A comparative study of HCFO-1224yd(Z), HCFO-1233zd(E) and HFO-1336mzz(Z) for high temperature heat pumps as potential replacements for R245fa was conducted by Mateu-Royo et al. (2019). Comparisons were performed in terms of coefficient of performance (COP) and Total Equivalent Warming Impact (TEWI) at the condensing and evaporating temperature ranges of 115°C to 145°C and 45 to 75°C, respectively. A single-stage cycle with and without internal heat exchanger (IHX) was used as a reference for the evaluations. COP comparisons indicate a 27% with HCFO-1233zd(E) as well as about 21 and 17% with HFO-1336mzz(Z) and HCFO-1224yd(Z) respectively over R245fa. In terms of TEWI, the reductions are between 59 and 61% with HCFO-1233zd(E) showing the highest effect, followed by HCFO-1224yd(Z) and HFO-1336mzz(Z).

Table2 highlights some representative investigations with several working fluids used in industrial heat pumps, heat pipes and ORC systems.

**Table 2. Fluid in industrial heat pumps, heat pipe and ORC systems**

| Working fluids | System | System performance | References |
|---|---|---|---|
| Butane, R123, R245fa, R600a, R245ca, R141b | Basic ORC | $\eta_{ex}$ : 25.1%, 23.8%, 25.4%, 26.1%, 24.2%, 23.4% | (Uusitalo, Honkatukia, and Turunen-Saaresti 2017) |
| Benzene, cyclohexane, Toluene, P-xylene | Basic ORC | $\eta_{th}$: 16.1%, 17.4, 16.3%, 15.9% | (Khatita et al. 2014) |
| Cyclohexane, Decane, Nonane, Octane, Heptane, Cyclopentane, Hexane, Isohexane | Basic ORC | $\eta_{th}$ : 18.02%, 17.64%, 17.57%, 17.12%, 16.8%, 16.64%, 14.4%, 13.62%, | (Shu et al. 2014) |
| R290, R600, R600a, R601, R601a, R134a, R227ea, R245fa | Basic ORC | $\eta_{th}$: 10.62%, 11.33%, 11.28%, 11.34%,11.17%, 10.54%, 10.16%, 11.36% | (Guo et al. 2015) |
| R245fa/R227ea(9/1), R245fa/R227ea(8/2), R245fa/R227ea(7/3), R245fa/R227ea(6/4) | Basic ORC | $\eta_{ex}$: 49.42%, 49.4%, 48.86%, 47.41% | (Feng et al. 2015) |
| Cyclohexane, benzene, toluene | ORC with preheater | $\eta_{ex}$: 27.11%, 25.62%, 18.04% | (Fergani, Touil, and Morosuk 2016) |
| Working fluids | System | System performance | References |
| Working fluids | System | System performance | References |
| R113, R245FA, R601, R601A | ORC with preheater | $\eta_{th}$: 19.69%, 19.42%, 19.39%, 19.46% | (Hærvig, Sørensen, and Condra 2016) |

**Table 2. (Continued)**

| Working fluids | System | System performance | References |
|---|---|---|---|
| R245fa, R600a, R245ca, R1234yf, R1233zd | ORC with preheater | $\eta_{th}$: 4.48%, 4.38%, 4.56%, 4.08%, 4.46% | (M. Yang and Yeh 2014) |
| R134a, R236fa, R245fa | micro combined heat and power (CHP) plant | $\eta_{th}$: 6.9%, 7.4%, 10.0 | (Tempesti, Manfrida, and Fiaschi 2012; Tempesti and Fiaschi 2013) |
| Water and brine | Geothermal, Space-Water Heating, ORC | $\eta_{th}$: 69.6% and $\eta ex$: 42.8% | (Ezzat and Dincer 2016) |
| Syngas $H_2$, $CO_2$ | Solar tower heated steam-based biomass gasification | $\eta_{th}$: 56.09% $\eta_{ex}$: 54.86%, | (Bai et al. 2015) |
| R410A | low-temperature WHR HP | COP: 5.5 | (Hervás-Blasco et al. 2019) |
| Propane | high temperature WHR HP | COP: 3.1 | (Bamigbetan et al. 2019) |
| R600 | HP - paper production process | COP: 4.4-5.1 | (Wemmers et al. 2017) |
| Mixtures: R290+R600, R600+R601 | high temperature HP | Improving COP by 13.5%. | (Opeyemi Bamigbetan et al. 2016) |
| R123, R141b, R245fa, R1234ze(Z) | HP for steam generation using geothermal water | COP: 3.55, 3.8, 3.22, 3.44 | (Hongjie Chen, Zhu, and Lu 2018) |
| R-744 | Supercritical $CO_2$ power cycle | $\eta_{th}$: 33% | (Olumayegun and Wang 2019) |
| R-744 | $CO_2$ transcritical power cycle | $\eta_{th}$: 9.2% | (Y. Chen et al. 2006) |
| LiBr-water | Absorp HP power plant | COP: 1.77 | (Z. Y. Xu et al. 2018) |
| LiBr-water | Absorp HP | COP: 1.65- 2.0 | (Keil et al. 2008) |
| $H_2O$-$NH_3$ | Hybrid absorp HP | COP: 5.49 $\eta_{ex}$ : 27.62%, | (Liu et al. 2018) |
| R134a-$TiO_2$ | Refrigeration systems | Energy saving: 26.1% | (Bi, Shi, and Zhang 2008) |
| R134a-$Al_2O_3$ | | COP increase 4.4% | (Jwo et al. 2009) |

**Table 2. (Continued)**

| Working fluids | System | System performance | References |
|---|---|---|---|
| R134a and R12-Carbon nanotube | Refrigeration systems | Heat transfer coefficient enhancement up to 36.6% | (Park and Jung 2007) |
| R134a-CuO | | COP: 5.75 | (Coumaressin and Palaniradja 2014) |
| $Al_2O_3$ nanoparticles | Cogeneration system in cement industry | Energy saving: 0.1% and 0.5% | (Hadi 2015) |
| Water | Heat pipe heat in air conditioning | $\varepsilon_{ff}$ : 48% | (F. Yang, Yuan, and Lin 2003) |
| R134a, R410A | Heat pipe | $\varepsilon_{ff}$ 35.6% and 57.7% | (Gedik, Yilmaz, and Kurt 2016) |
| Steam | Cement industries | $\eta_{th}$ :84% and $\eta_{ex}$ : 48% | (Sogut, Oktay, and Hepbasli 2009) |
| Methanol | HPHE for surgery rooms in hospitals | $\eta_{th}$ : 16% | (Noie-Baghban and Majideian 2000) |
| Water | Heat pipe | $\eta_{th}$ :33.3% | (W. Wang et al. 2019) |
| Water | Heat pipe in coke oven gas | $\eta_{th}$ :69% | (S. Zhang 2019b) |

$\eta_{th}$: thermal efficiency, $\eta_{ex}$: exergy, $\varepsilon_{ff}$: effectiveness.

## CONCLUSION

Waste heat is abundantly generated in a wide range of temperatures in building and industrial processes and continues to be lost in large amounts to the environment but research devotes considerable efforts to maximize recovery of this important energy source, particularly in the current environmental and economic context. This is achieved by existing equipment as well as the development of new approaches and technologies.

This chapter provides a comprehensive overview of the waste heat recovery potential, the available technologies, and those in development for its harnessing, with their advantages and challenges. It is a detailed survey of waste heat sources organized in sections for heat characterization, availability, assessment methods, recovery equipment characteristics, as well as compatible working fluids. Particular attention is given to technologies having the ability to recover waste heat over a wide range of high to low temperatures, where waste heat is more suitable for power generation (Rankine, ORC, Kalina cycles and other variations). Heating and cooling technologies such as industrial heat pump types for cooling, heat upgrading including mechanical or thermal recompression, passive heat displacement devices such as heat exchangers and heat pipes are also discussed.

The main concluding elements emerging from this assessment may be stated as follows:

- Considering the prevailing context of increasing energy consumption and growing awareness of climate change, waste heat recovery has become significant research efforts devoted to the developpment of this activity. Several research aspects such as waste heat characterization (type, quality, quantity, etc…), use of conventional/unconventional technologies, development of new concepts, economic considerations, use of new methodologies and approaches for optimised waste heat estimation and recovery, where source-sink compatibility, temporal availability and adapted technology selection continue to be proposed and adopted.
- While some of WHR barriers are process or application specific, many are crosscutting across several heat recovery approaches. They generally reduce the efficiency of adopted heat recovery systems and in some cases may prevent heat recovery from being installed. This is mainly due to the application's specific constraints, and the nature of key restrictions (use, design, thermodynamics, cost, heat stream quality and composition, temperature, process and applications).
- Several existing methods commonly used to estimate WHR potential are very useful and applicable in many cases of interest. However, the choice of any method depends upon a number of site specific factors (i.e., location, waste heat potential, costs, etc…), in

addition to more technical considerations already mentioned (fluid nature, temperature, composition etc…).

- Given the environmental and economic contexts, a new tendency in modern waste heat recovery technology is related to the use of natural refrigerants (hydrocarbons, carbon dioxide, water and ammonia). However, some challenges remain in their development, use and availability of adapted equipment.
- Working fluids for heat recovery systems have an important impact on the performance and the environmental footprint of systems. The selected fluid should correspond to an acceptable compromise on these aspects. For this reason, a great deal of research continues to investigate and develop new fluids to cover wider ranges of WHR conditions.

## Acknowledgments

The authors would like to acknowledge the support of CanmetENERGY-Varennes (Natural Resources Canada), and Energy Technologies Laboratory (Hydro-Quebec).

## References

Abadi, Bamorovat Gholamreza, and Kyung Chun Kim. 2017. "Investigation of Organic Rankine Cycles with Zeotropic Mixtures as a Working Fluid: Advantages and Issues." *Renewable and Sustainable Energy Reviews* 73 (January). Elsevier Ltd: 1000–1013. doi: 10.1016/j.rser.2017.02.020.

Agathokleous, Rafaela, Giuseppe Bianchi, Gregoris Panayiotou, Lazaros Arestia, Maria C. Argyrou, Giorgos S. Georgiou, Savvas A. Tassou, et al. 2019. "Waste Heat Recovery in the EU Industry and Proposed New Technologies." *Energy Procedia* 161. Elsevier B.V.: 489–96. doi: 10.1016/j.egypro.2019.02.064.

Aidoun, Z., K. Ameur, M. Falsafioon, and M. Badache. 2019. "Current Advances in Ejector Modeling, Experimentation and Applications for Refrigeration and Heat Pumps. Part 1: Single-Phase Ejectors." *Inventions* 4 (15): 1–73. doi: 10.3390/inventions4010015.

Al-Rabghi, O. M., M. Akyurt, Y. S. H. Najjar, and T. Alp. 1993. "Heat Exchangers for Waste-Heat Recovery." *Energy & Environment* 4 (3): 284–306. doi: 10.1177/0958305x9300400305.

Arpagaus, Cordin, Frédéric Bless, Jürg Schiffmann, and Stefan S. Bertsch. 2016. "Multi-Temperature Heat Pumps: A Literature Review." *International Journal of Refrigeration* 69. Elsevier Ltd: 437–65. doi: 10. 1016/j.ijrefrig.2016.05.014.

Arpagaus, Cordin, Frédéric Bless, Michael Uhlmann, Jürg Schiffmann, and Stefan S. Bertsch. 2018. "High Temperature Heat Pumps: Market Overview, State of the Art, Research Status, Refrigerants, and Application Potentials." In *17th International Refrigeration and Air Conditioning Conference*, 1–10. Purdue University. Paper 1876. doi: 10. 1016/j.energy.2018.03.166.

ASHRAE. 1965. "Steam-Jet Refrigeration Equipment." In *Guide and Data Book*, 655–60.

Badache, Messaoud, Zine Aidoun, Parham Eslami-Nejad, and Daniela Blessent. 2019. *Ground-Coupled Natural Circulating Devices (Thermosiphons): A Review of Modeling, Experimental and Development Studies. Inventions.* Vol. 4. doi: 10.3390/inventions 4010014.

Badache, Messaoud, Mohamed Ouzzane, Parham Eslami-nejad, and Zine Aidoun. 2018. "Experimental Study of a Carbon Dioxide Direct-Expansion Ground Source Heat Pump (CO2-DX-GSHP)." *Applied Thermal Engineering* 130: 1480–88. doi: 10.1016/j.applthermaleng. 2017.10.159.

Bai, Zhang, Qibin Liu, Jing Lei, Hongqiang Li, and Hongguang Jin. 2015. "A Polygeneration System for the Methanol Production and the Power Generation with the Solar – Biomass Thermal Gasification." *Energy Conversion and Management* 102. Elsevier Ltd: 190–201. doi: 10.1016/ j.enconman.2015.02.031.

Bamigbetan, Opeyemi, T. M. Eikevik, P. Nekså, M. Bantle, and C. Schlemminger. 2019. "The Development of a Hydrocarbon High Temperature Heat Pump for Waste Heat Recovery." *Energy* 173. Elsevier Ltd: 1141–53. doi: 10.1016/j.energy.2019.02.159.

Bamigbetan, Opeyemi, Trygve M. Eikevik, Petter Nekså, and Michael Bantle. 2017. "Review of Vapour Compression Heat Pumps for High Temperature Heating Using Natural Working Fluid." *International Journal of Refrigeration* 80. Elsevier Ltd: 197–211. doi: 10.1016/j.ijrefrig.2017.04.021.

Bamigbetan, Opeyemi, Trygve Eikevik, Petter Nekså, and Michael Bantle. 2016. "Evaluation of Natural Working Fluids for the Development of High Temperature Heat Pumps." In *12th IIR Gustav Lorentzen Natural Working Fluids Conference*. Edinburgh, Scotland. Paper 1092. doi: 10.18462/iir.gl.2016.1092.

Bao, Junjiang, and Li Zhao. 2013. "A Review of Working Fluid and Expander Selections for Organic Rankine Cycle." *Renewable and Sustainable Energy Reviews* 24. Elsevier: 325–42. doi: 10.1016/j.rser.2013.03.040.

Bi, Sheng-shan, Lin Shi, and Li-li Zhang. 2008. "Application of Nanoparticles in Domestic Refrigerators." *Applied Thermal Engineering* 28: 1834–43. doi: 10.1016/j.applthermaleng.2007.11.018.

Borgnakke, C., and R. E. Sonntag. 2012. *Fundamentals of Thermodynamics*. Edited by Wiley. 8th ed. doi: 10.1201/b12725-3.

Campana, F., M. Bianchi, L. Branchini, A. De Pascale, A. Peretto, M. Baresi, A. Fermi, N. Rossetti, and R. Vescovo. 2013. "ORC Waste Heat Recovery in European Energy Intensive Industries: Energy and GHG Savings." *Energy Conversion and Management* 76. Elsevier Ltd: 244–52. doi: 10.1016/j.enconman.2013.07.041.

Carcasci, Carlo, Riccardo Ferraro, and Edoardo Miliotti. 2014. "Thermodynamic Analysis of an Organic Rankine Cycle for Waste Heat Recovery from Gas Turbines." *Energy* 65. Elsevier Ltd: 91–100. doi: 10.1016/j.energy.2013.11.080.

Chen, Hongjie, Jiahua Zhu, and Wei Lu. 2018. "Optimized Selection of One- and Two-Stage Ejectors under Design and off-Design Conditions."

*Energy Conversion and Management* 173 (June). Elsevier: 743–52. doi: 10.1016/j.enconman.2018.06.113.

Chen, Huijuan, D. Yogi Goswami, and Elias K. Stefanakos. 2010. "A Review of Thermodynamic Cycles and Working Fluids for the Conversion of Low-Grade Heat." *Renewable and Sustainable Energy Reviews* 14 (9). Elsevier Ltd: 3059–67. doi: 10.1016/j.rser.2010.07.006.

Chen, Y., P. Lundqvist, A. Johansson, and P. Platell. 2006. "A Comparative Study of the Carbon Dioxide Transcritical Power Cycle Compared with an Organic Rankine Cycle with R123 as Working Fluid in Waste Heat Recovery." *Applied Thermal Engineering* 26 (17–18): 2142–47. doi: 10.1016/j.applthermaleng.2006.04.009.

Chintala, Venkateswarlu, Suresh Kumar, and Jitendra K. Pandey. 2018. "A Technical Review on Waste Heat Recovery from Compression Ignition Engines Using Organic Rankine Cycle." *Renewable and Sustainable Energy Reviews* 81 (June 2016). Elsevier Ltd: 493–509. doi: 10.1016/j.rser.2017.08.016.

Chua, K. J., S. K. Chou, and W. M. Yang. 2010. "Advances in Heat Pump Systems: A Review." *Applied Energy* 87 (12). Elsevier Ltd: 3611–24. doi: 10.1016/j.apenergy.2010.06.014.

Coumaressin, T, and K Palaniradja. 2014. "Performance Analysis of a Refrigeration System Using Nano Fluid." *International Journal of Advanced Mechanical Engineering* 4 (4): 459–70.

Cuce, P.M., and S. Riffat. 2015. "A Comprehensive Review of Heat Recovery Systems for Building Applications." *Renewable and Sustainable Energy Reviews* 47. Elsevier: 665–82. doi: 10.1016/j.rser.2015.03.087.

Dai, Xiaoye, Lin Shi, and Weizhong Qian. 2019. "Review of the Working Fluid Thermal Stability for Organic Rankine Cycles." *Journal of Thermal Science* 28 (4): 597–607. doi: 10.1007/s11630-019-1119-3.

Escalante, E. S. R., J. A. de Carvalho Junior, and J. A. P. Balestieri. 2017. "Effect of Working Fluids on Organic Rankine Cycle for the Recovery of Low-Grade Waste Heat." In *The 12th Latin-Americain Congress on Electricity Generation and Transmission*, 1–6. Mar del Plata, Argentina. doi: 10.1016/j.energy.2004.01.004.

Eyerer, Sebastian, Fabian Dawo, Johannes Kaindl, Christoph Wieland, and Hartmut Spliethoff. 2019. "Experimental Investigation of Modern ORC Working Fluids R1224yd(Z) and R1233zd(E) as Replacements for R245fa." *Applied Energy* 240 (February). Elsevier: 946–63. doi: 10.1016/j.apenergy.2019.02.086.

Ezzat, M. F., and I. Dincer. 2016. "ScienceDirect Energy and Exergy Analyses of a New Geothermal – Solar Energy Based System." *Solar Energy* 134. Elsevier Ltd: 95–106. doi: 10.1016/j.solener.2016.04.029.

Fakeye, B. A., and S. O. Oyedepo. 2018. "A Review of Working Fluids for Organic Rankine Cycle (ORC) Applications." In *IOP Conference Series: Materials Science and Engineering*. Vol. 413. doi: 10.1088/1757-899X/413/1/012019.

Feng, Yongqiang, Tzu Chen Hung, Kowalski Greg, Yaning Zhang, Bingxi Li, and Jinfu Yang. 2015. "Thermoeconomic Comparison between Pure and Mixture Working Fluids of Organic Rankine Cycles (ORCs) for Low Temperature Waste Heat Recovery." *Energy Conversion and Management* 106. Elsevier Ltd: 859–72. doi: 10.1016/j.enconman.2015.09.042.

Fergani, Zineb, Djamel Touil, and Tatiana Morosuk. 2016. "Multi-Criteria Exergy Based Optimization of an Organic Rankine Cycle for Waste Heat Recovery in the Cement Industry." *Energy Conversion and Management* 112. Elsevier Ltd: 81–90. doi: 10.1016/j.enconman.2015.12.083.

Firth, Anton, Bo Zhang, and Aidong Yang. 2019. "Quantification of Global Waste Heat and Its Environmental Effects." *Applied Energy* 235 (April 2018): 1314–34. doi: 10.1016/j.apenergy.2018.10.102.

Forman, Clemens, Ibrahim Kolawole Muritala, Robert Pardemann, and Bernd Meyer. 2016. "Estimating the Global Waste Heat Potential." *Renewable and Sustainable Energy Reviews* 57. Elsevier: 1568–79. doi: 10.1016/j.rser.2015.12.192.

Galanis, N., E. Cayer, P. Roy, E. S. Denis, and M. Désilets. 2009. "Electricity Generation from Low Temperature Sources." *Journal of Applied Fluid Mechanics* 2 (2): 55–67.

Gedik, Engin, Mustafa Yilmaz, and Hüseyin Kurt. 2016. "Experimental Investigation on the Thermal Performance of Heat Recovery System with Gravity Assisted Heat Pipe Charged with R134a and R410A." *Applied Thermal Engineering* 99. Elsevier Ltd: 334–42. doi: 10.1016/j.applthermaleng.2015.12.075.

Guo, Cong, Xiaoze Du, Lijun Yang, and Yongping Yang. 2015. "Organic Rankine Cycle for Power Recovery of Exhaust Flue Gas." *Applied Thermal Engineering* 75. Elsevier Ltd: 135–44. doi: 10.1016/j.applthermaleng.2014.09.080.

Haddad, Cynthia, Christelle Périlhon, Amélie Danlos, Maurice Xavier François, and Georges Descombes. 2014. "Some Efficient Solutions to Recover Low and Medium Waste Heat: Competitiveness of the Thermoacoustic Technology." *Energy Procedia* 50. Elsevier B.V.: 1056–69. doi: 10.1016/j.egypro.2014.06.125.

Hadi, Ali Amer A. 2015. "Nanoparticles Concentration and Environmental Effects on Cogeneration System in Cement Industry." *International Journal of Engineering Research and General Science* 3 (5): 786–98.

Hærvig, J., K. Sørensen, and T. J. Condra. 2016. "Guidelines for Optimal Selection of Working Fluid for an Organic Rankine Cycle in Relation to Waste Heat Recovery." *Energy* 96. Elsevier Ltd: 592–602. doi: 10.1016/j.energy.2015.12.098.

Hervás-Blasco, Estefanía, Emilio Navarro-Peris, Francisco Barceló-Ruescas, and José Miguel Corberán. 2019. "Improved Water to Water Heat Pump Design for Low-Temperature Waste Heat Recovery Based on Subcooling Control." *International Journal of Refrigeration* 106: 374–83. doi: 10.1016/j.ijrefrig.2019.06.030.

Hesselgreaves, J. E. 2001. *Compact Heat Exchangers, Selections, Design and Operation*. Elsevier Science & Technology Books.

IEA. 1995. "*Industrial Heat Pumps. Experiences, Potential and Global Environmental Benefits. IEA Annex 21.*" Sittard, the Netherlands.

IEO2019. 2019. "*International Energy Outlook 2019, with Projections to 2050.*" doi: 10.1080/01636609609550217.

Jouhara, Hussam, Navid Khordehgah, Sulaiman Almahmoud, Bertrand Delpech, Amisha Chauhan, and Savvas A. Tassou. 2018. "Waste Heat

Recovery Technologies and Applications." *Thermal Science and Engineering Progress* 6. Elsevier: 268–89. doi: 10.1016/j.tsep.2018.04.017.

Jouhara, Hussam, and Abdul Ghani Olabi. 2018. "Editorial: Industrial Waste Heat Recovery." *Energy* 160. Elsevier Ltd: 1–2. doi: 10.1016/j.energy.2018.07.013.

Jwo, C. S., L. Y. Jeng, T. P. Teng, and H. Chang. 2009. "Effects of Nanolubricant on Performance of Hydrocarbon Refrigerant System." *Journal of Vacuum Science & Technology B: Microelectronics and Nanometer Structures Processing, Measurement, and Phenomena* 27 (3): 1473–77.

Keil, Christian, Stefan Plura, Michael Radspieler, and Christian Schweigler. 2008. "Application of Customized Absorption Heat Pumps for Utilization of Low-Grade Heat Sources." *Applied Thermal Engineering* 28 (16): 2070–76. doi: 10.1016/j.applthermaleng.2008.04.012.

Khatita, Mohammed A., Tamer S. Ahmed, Fatma H. Ashour, and Ibrahim M. Ismail. 2014. "Power Generation Using Waste Heat Recovery by Organic Rankine Cycle in Oil and Gas Sector in Egypt: A Case Study." *Energy* 64. Elsevier Ltd: 462–72. doi: 10.1016/j.energy.2013.11.011.

Kolasinska, Ewan, and Piotr Kolasinska. 2016. "A Review on Electroactive Polymers for Waste Heat Recovery." *Materials* 9 (485): 1–15. doi: 10.3390/ma9060485.

Landelle, Arnaud, Nicolas Tauveron, Philippe Haberschill, Rémi Revellin, and Stéphane Colasson. 2017. "Organic Rankine Cycle Design and Performance Comparison Based on Experimental Database." *Applied Energy* 204. Elsevier Ltd: 1172–87. doi: 10.1016/j.apenergy.2017.04.012.

Lecompte, Steven, Henk Huisseune, Martijn Van Den Broek, Bruno Vanslambrouck, and Michel De Paepe. 2015. "Review of Organic Rankine Cycle (ORC) Architectures for Waste Heat Recovery." *Renewable and Sustainable Energy Reviews* 47. Elsevier: 448–61. doi: 10.1016/j.rser.2015.03.089.

Lemmens, Sanne. 2015. "Waste Heat Recovery : Potential, Policy Challenges and Technology Choices." In *38th IAEE International Conference*. Antalya, Turkey.

Li, Haoran, and Natasa Nord. 2018. "Transition to the 4th Generation District Heating - Possibilities, Bottlenecks, and Challenges." *Energy Procedia* 149. Elsevier B.V.: 483–98. doi: 10.1016/j.egypro.2018.08.213.

Liu, Changchun, Yingchun Jiang, Wei Han, and Qilan Kang. 2018. "A High-Temperature Hybrid Absorption-Compression Heat Pump for Waste Heat Recovery." *Energy Conversion and Management* 172 (July). Elsevier: 391–401. doi: 10.1016/j.enconman.2018.07.027.

Lund, Rasmus, and Urban Persson. 2016. "Mapping of Potential Heat Sources for Heat Pumps for District Heating in Denmark." *Energy* 110. Elsevier Ltd: 129–38. doi: 10.1016/j.energy.2015.12.127.

Mahmoudi, A., M. Fazli, and M. R. Morad. 2018. "A Recent Review of Waste Heat Recovery by Organic Rankine Cycle." *Applied Thermal Engineering* 143 (July). Elsevier: 660–75. doi: 10.1016/j.applthermaleng.2018.07.136.

Maraver, Daniel, Javier Royo, Vincent Lemort, and Sylvain Quoilin. 2014. "Systematic Optimization of Subcritical and Transcritical Organic Rankine Cycles (ORCs) Constrained by Technical Parameters in Multiple Applications." *Applied Energy* 117. Elsevier Ltd: 11–29. doi: 10.1016/j.apenergy.2013.11.076.

Mardiana-Idayu, A., and S. B. Riffat. 2012. "Review on Heat Recovery Technologies for Building Applications." *Renewable and Sustainable Energy Reviews* 16 (2). Elsevier Ltd: 1241–55. doi: 10.1016/j.rser.2011.09.026.

Mateu-Royo, Carlos, Joaquín Navarro-Esbrí, Adrián Mota-Babiloni, Marta Amat-Albuixech, and Francisco Molés. 2019. "Thermodynamic Analysis of Low GWP Alternatives to HFC-245fa in High-Temperature Heat Pumps: HCFO-1224yd(Z), HCFO-1233zd(E) and HFO-1336mzz(Z)." *Applied Thermal Engineering* 152 (January). Elsevier: 762–77. doi: 10.1016/j.applthermaleng.2019.02.047.

Miró, Laia, Sarah Brückner, and Luisa F. Cabeza. 2015. "Mapping and Discussing Industrial Waste Heat (IWH) Potentials for Different Countries." *Renewable and Sustainable Energy Reviews* 51. Elsevier: 847–55. doi: 10.1016/j.rser.2015.06.035.

Mondejar, M. E., J. G. Andreasen, L. Pierobon, U. Larsen, M. Thern, and F. Haglind. 2018. "A Review of the Use of Organic Rankine Cycle Power Systems for Maritime Applications." *Renewable and Sustainable Energy Reviews* 91. Elsevier Ltd: 126–51. doi: 10.1016/j.rser.2018.03.074.

Nasir, Muhammad Tauseef, and Kyung Chun Kim. 2016. "Working Fluids Selection and Parametric Optimization of an Organic Rankine Cycle Coupled Vapor Compression Cycle (ORC-VCC) for Air Conditioning Using Low Grade Heat." *Energy and Buildings* 129. Elsevier B.V.: 378–95. doi: 10.1016/j.enbuild.2016.07.068.

Nellissen, Philippe, and Stefan Wolf. 2015. "Heat Pumps in Non-Domestic Applications in Europe: Potential for an Energy Revolution." In *8th EHPA European Heat Pump Forum*, 1–17. Brussels.

Noie-Baghban, S. H., and G. R. Majideian. 2000. "Waste Heat Recovery Using Heat Pipe Heat Exchanger (HPHE) for Surgery Rooms in Hospitals." *Applied Thermal Engineering* 20 (14): 1271–82. doi: 10.1016/S1359-4311(99)00092-7.

O'Rielly, Kristine, and Jack Jeswiet. 2015. "Improving Industrial Energy Efficiency through the Implementation of Waste Heat Recovery Systems." *Transactions of the Canadian Society for Mechanical Engineering* 39 (1): 125–36. doi: 10.1139/tcsme-2015-0010.

Oldershaw, J., G. Morrell, J. Besseling, and S. Slater. 2016. "*Barriers and Enablers to Recovering Surplus Heat in Industry*." London, UK.

Olumayegun, Olumide, and Meihong Wang. 2019. "Dynamic Modelling and Control of Supercritical CO2 Power Cycle Using Waste Heat from Industrial Processes." *Fuel* 249. Elsevier: 89–102. doi: 10.1016/j.fuel.2019.03.078.

Panwar, N. L., S. C. Kaushik, and Surendra Kothari. 2011. "Role of Renewable Energy Sources in Environmental Protection: A Review."

*Renewable and Sustainable Energy Reviews* 15 (3). Elsevier Ltd: 1513–24. doi: 10.1016/j.rser.2010.11.037.

Papadopoulos, Athanasios I., Mirko Stijepovic, and Patrick Linke. 2010. "On the Systematic Design and Selection of Optimal Working Fluids for Organic Rankine Cycles." *Applied Thermal Engineering* 30 (6–7). Elsevier Ltd: 760–69. doi: 10.1016/j.applthermaleng.2009.12.006.

Papapetrou, Michael, George Kosmadakis, Andrea Cipollina, Umberto La Commare, and Giorgio Micale. 2018. "Industrial Waste Heat: Estimation of the Technically Available Resource in the EU per Industrial Sector, Temperature Level and Country." *Applied Thermal Engineering* 138 (April). Elsevier: 207–16. doi: 10.1016/j.applthermaleng.2018.04.043.

Park, Ki-jung, and Dongsoo Jung. 2007. "Boiling Heat Transfer Enhancement with Carbon Nanotubes for Refrigerants Used in Building Air-Conditioning." *Energy & Buildings* 39: 1061–64. doi: 10.1016/j.enbuild.2006.12.001.

Pelligrino, J. L., N. Margolis, M. Justianiano, M. Miller, and A. Thedki. 2004. "*Energy Use, Loss and Opportunities Analysis: U.S. Manufactring & Mining*."

Pierobon, L., F. Haglind, R. Kandepu, A. Fermi, and N. Rossetti. 2013. "Technologies for Waste Heat Recovery in Off-Shore Applications." In *ASME 2013 International Mechanical Engineering Congress and Exposition*, 1–10. San Diego, USA.

Quoilin, Sylvain, Martijn Van Den Broek, Sébastien Declaye, Pierre Dewallef, and Vincent Lemort. 2013. "Techno-Economic Survey of Organic Rankine Cycle (ORC) Systems." *Renewable and Sustainable Energy Reviews* 22: 168–86. doi: 10.1016/j.rser.2013.01.028.

Reay, David. 1999. *Learning from Experiences with, Compact Heat Exchangers*. Netherlands: CADDET Energy Efficiency.

Reddy, C. C. S., S. V. Naidu, and G. P. Rangaiah. 2013. "Waste Heat Recovery Methods and Technologies." *Chemical Engineering*, 1–16.

Saidur, R., S. N. Kazi, M. S. Hossain, M. M. Rahman, and H. A. Mohammed. 2011. "A Review on the Performance of Nanoparticles Suspended with Refrigerants and Lubricating Oils in Refrigeration

Systems." *Renewable and Sustainable Energy Reviews* 15 (1): 310–23. doi: 10.1016/j.rser.2010.08.018.

Sarkar, Jahar. 2015. "Review and Future Trends of Supercritical CO2 Rankine Cycle for Low-Grade Heat Conversion." *Renewable and Sustainable Energy Reviews* 48. Elsevier: 434–51. doi: 10.1016/j.rser.2015.04.039.

Shengjun, Zhang, Wang Huaixin, and Guo Tao. 2011. "Performance Comparison and Parametric Optimization of Subcritical Organic Rankine Cycle (ORC) and Transcritical Power Cycle System for Low-Temperature Geothermal Power Generation." *Applied Energy* 88 (8). Elsevier Ltd: 2740–54. doi: 10.1016/j.apenergy.2011.02.034.

Shu, Gequn, Xiaoning Li, Hua Tian, Xingyu Liang, Haiqiao Wei, and Xu Wang. 2014. "Alkanes as Working Fluids for High-Temperature Exhaust Heat Recovery of Diesel Engine Using Organic Rankine Cycle." *Applied Energy* 119. Elsevier Ltd: 204–17. doi: 10.1016/j.apenergy.2013.12.056.

Simeoni, Patrizia, Gellio Ciotti, Mattia Cottes, and Antonella Meneghetti. 2019. "Integrating Industrial Waste Heat Recovery into Sustainable Smart Energy Systems." *Energy* 175. Elsevier Ltd: 941–51. doi: 10.1016/j.energy.2019.03.104.

Sogut, M. Z., Z. Oktay, and A. Hepbasli. 2009. "Energetic and Exergetic Assessment of a Trass Mill Process in a Cement Plant." *Energy Conversion and Management* 50 (9). Elsevier Ltd: 2316–23. doi: 10.1016/j.enconman.2009.05.013.

Sprouse, Charles, and Christopher Depcik. 2013. "Review of Organic Rankine Cycles for Internal Combustion Engine Exhaust Waste Heat Recovery." *Applied Thermal Engineering* 51 (1–2). Elsevier Ltd: 711–22. doi: 10.1016/j.applthermaleng.2012.10.017.

Srimuang, W., and P. Amatachaya. 2012. "A Review of the Applications of Heat Pipe Heat Exchangers for Heat Recovery." *Renewable and Sustainable Energy Reviews* 16 (6). Elsevier Ltd: 4303–15. doi: 10.1016/j.rser.2012.03.030.

Tchanche, Bertrand F., Gr Lambrinos, A. Frangoudakis, and G. Papadakis. 2011. "Low-Grade Heat Conversion into Power Using Organic Rankine

Cycles - A Review of Various Applications." *Renewable and Sustainable Energy Reviews* 15 (8). Elsevier Ltd: 3963–79. doi: 10. 1016/j.rser.2011.07.024.

Tempesti, Duccio, and Daniele Fiaschi. 2013. "Thermo-Economic Assessment of a Micro CHP System Fuelled by Geothermal and Solar Energy." *Energy* 58. Elsevier Ltd: 45–51. doi: 10.1016/j.energy.2013. 01.058.

Tempesti, Duccio, Giampaolo Manfrida, and Daniele Fiaschi. 2012. "Thermodynamic Analysis of Two Micro CHP Systems Operating with Geothermal and Solar Energy." *Applied Energy* 97. Elsevier Ltd: 609–17. doi: 10.1016/j.apenergy.2012.02.012.

Thekdi, Arvind, and Sachin U. Nimbalkar. 2014. "Industrial Waste Heat Recovery: Potential Applications, Available Technologies and Crosscutting R&D Opportunities." *Oak Ridge National Laboratory*. Oak Ridge. doi: 10.2172/1185778.

Tumanovskii, A. G., A. L. Shvarts, E. V. Somova, E. Kh Verbovetskii, G. D. Avrutskii, S. V. Ermakova, R. N. Kalugin, and M. V. Lazarev. 2017. "Review of the Coal-Fired, over-Supercritical and Ultra-Supercritical Steam Power Plants." *Thermal Engineering* 64 (2): 83–96. doi: 10.1134/ S0040601517020082.

USDOE. 2008. "*Waste Heat Recovery: Technology and Opportunities in U.S. Industry*."

Uusitalo, Antti, Juha Honkatukia, and Teemu Turunen-Saaresti. 2017. "Evaluation of a Small-Scale Waste Heat Recovery Organic Rankine Cycle." *Applied Energy* 192. Elsevier Ltd: 146–58. doi: 10.1016/j. apenergy.2017.01.088.

Van de Bor, D. M., C. A. Infante Ferreira, and Anton A. Kiss. 2015. "Low Grade Waste Heat Recovery Using Heat Pumps and Power Cycles." *Energy* 89. Elsevier Ltd: 864–73. doi: 10.1016/j.energy.2015.06.030.

Vance, David, Sachin Nimbalkar, Arvind Thekdi, Kristina Armstrong, Thomas Wenning, Joseph Cresko, and Mingzhou Jin. 2019. "Estimation of and Barriers to Waste Heat Recovery from Harsh Environments in Industrial Processes." *Journal of Cleaner Production* 222. Elsevier Ltd: 539–49. doi: 10.1016/j.jclepro.2019.03.011.

Varga, Zoltán, and Balázs Palotai. 2017. "Comparison of Low Temperature Waste Heat Recovery Methods." *Energy* 137: 1286–92. doi: 10.1016/j.energy.2017.07.003.

Vasiliev, Leonard, Leonid p Grakovich, Mikhail Rabetsky, and Leonid Jr. Vasiliev. 2013. "Heat Transfer Enhancement in Heat Pipes and Thermosyphons Using Nanotechnologies." *Heat Pipe Science and Technology, An International Journal* 4 (4): 251–75. doi: 10.1615/HeatPipeScieTech.2014011995.

Vélez, Fredy, José J. Segovia, M. Carmen Martín, Gregorio Antolín, Farid Chejne, and Ana Quijano. 2012. "A Technical, Economical and Market Review of Organic Rankine Cycles for the Conversion of Low-Grade Heat for Power Generation." *Renewable and Sustainable Energy Reviews* 16 (6). Elsevier Ltd: 4175–89. doi: 10.1016/j.rser.2012.03.022.

Wang, Wei-wei, Yang Cai, Run-zhe Liu, Fu-yun Zhao, and Di Liu. 2019. "Experimental and Numerical Investigations of a Radial Heat Pipe for Waste Heat Recovery." *Applied Thermal Engineering* 154 (December 2018). Elsevier: 602–13. doi: 10.1016/j.applthermaleng.2019.03.063.

Wang, Xuan, Ming Jin, Wei Feng, Gequn Shu, Hua Tian, and Youcai Liang. 2018. "Cascade Energy Optimization for Waste Heat Recovery in Distributed Energy Systems." *Applied Energy* 230 (June). Elsevier: 679–95. doi: 10.1016/j.apenergy.2018.08.124.

Wemmers, A K, A W M B Haasteren, P Kremers, and P an der Kamp. 2017. "Test Results R600 Pilot Heat Pump." In *12th IEA Heat Pump Conference*, 1–9. Rotterdam, the Netherlands.

Woolley, Elliot, Yang Luo, and Alessandro Simeone. 2018. "Industrial Waste Heat Recovery: A Systematic Approach." *Sustainable Energy Technologies and Assessments* 29 (June). Elsevier: 50–59. doi: 10.1016/j.seta.2018.07.001.

Xu, Qi, Saffa Riffat, and Shihao Zhang. 2019. "Review of Heat Recovery Technologies for Building Applications." *Energies* 12: 1–22. doi: 10.3390/en12071285.

Xu, Z. Y., H. C. Mao, D. S. Liu, and R. Z. Wang. 2018. "Waste Heat Recovery of Power Plant with Large Scale Serial Absorption Heat Pumps." *Energy* 165: 1097–1105. doi: 10.1016/j.energy.2018.10.052.

Xu, Z. Y., R. Z. Wang, and Chun Yang. 2019. "Perspectives for Low-Temperature Waste Heat Recovery." *Energy* 176: 1037–43. doi: 10.1016/j.energy.2019.04.001.

Xu, Zhenyuan, and Ruzhu Wang. 2017. "Absorption Heat Pump for Waste Heat Reuse: Current States and Future Development." *Frontiers in Energy* 11 (4): 414–36. doi: 10.1007/s11708-017-0507-1.

Yang, Feng, Xiugan Yuan, and Guiping Lin. 2003. "Waste Heat Recovery Using Heat Pipe Heat Exchanger for Heating Automobile Using Exhaust Gas." *Applied Thermal Engineering* 23 (3): 367–72. doi: 10.1016/S1359-4311(02)00190-4.

Yang, Min-hsiung, and Rong-hua Yeh. 2014. "Analyzing the Optimization of an Organic Rankine Cycle System for Recovering Waste Heat from a Large Marine Engine Containing a Cooling Water System." *Energy Conversion and Management* 88. Elsevier Ltd: 999–1010. doi: 10.1016/j.enconman.2014.09.044.

Zhang, Hainan, Shuangquan Shao, Changqing Tian, and Kunzhu Zhang. 2018. "A Review on Thermosyphon and Its Integrated System with Vapor Compression for Free Cooling of Data Centers." *Renewable and Sustainable Energy Reviews* 81. Elsevier Ltd: 789–98. doi: 10.1016/j.rser.2017.08.011.

Zhang, Shaojie. 2019a. "Pilot Study on Ascension-Pipe Heat Exchanger Used for Waste Heat Pilot Study on Heat Exchanger Used for Waste Heat Recovery of Coke Oven Gas Recovery of Coke Oven Gas." *Energy Procedia* 158. Elsevier B.V.: 26–31. doi: 10.1016/j.egypro.2019.01.029.

Zhang, Shaojie. 2019b. "Pilot Study on Ascension-Pipe Heat Exchanger Used for Waste Heat Recovery of Coke Oven Gas." *Energy Procedia* 158. Elsevier B.V.: 26–31. doi: 10.1016/j.egypro.2019.01.029.

Zhang, Yan, Yufeng Zhang, Xiaohui Yu, Jing Guo, Na Deng, Shengming Dong, Zhonglu He, and Xuelian Ma. 2017. "Analysis of a High Temperature Heat Pump Using BY-5 as Refrigerant." *Applied Thermal Engineering* 127. Elsevier Ltd: 1461–68. doi: 10.1016/j.applthermaleng.2017.08.072.

Ziviani, Davide, Asfaw Beyene, and Mauro Venturini. 2014. “Advances and Challenges in ORC Systems Modeling for Low Grade Thermal Energy Recovery.” *Applied Energy* 121. Elsevier Ltd: 79–95. doi: 10.1016/j.apenergy.2014.01.074.

In: Waste Heat: Sources and Uses
Editor: Michèle Picard
ISBN: 978-1-53617-355-0


*Chapter 4*

# WASTE HEAT RECOVERY FOR THE CEMENT SECTOR

***Vahab Ghalandari[1,*,†], Hamidreza Bagheri[2,†] and Ali Mohebbi[2]***

[1]Kerman Momtazan Cement Company, Kerman, Iran
[2]Department of Chemical Engineering, Faculty of Engineering, Shahid Bahonar University of Kerman, Kerman, Iran

## ABSTRACT

The cement sector is one of the most-energy using industries around the world. Generally, 75% of the used energy in the cement sector is thermal energy, and 25% is electrical energy. The considerable part of thermal energy, which is consumed to produce clinker in cement plants, is wasted from stacks in the form of convection. A waste heat recovery (WHR) system would be utilized to raise the efficiency of a cement plant; hence, it could reduce greenhouse gas emissions. Besides, it could decrease the exhaust gas temperature and the amount of waste heat to the atmosphere. In this chapter, first, the different types of WHR technologies

---

* Corresponding Author's Email: ghalandari.v@gmail.com.
† The authors shared the first authorship and contributed equally to this chapter.

that are used in cement plants to produce electricity were reviewed. Then, the potential for electricity production from heat exhaust gases of three cement plants by using WHR technology was discussed. Finally, the economic feasibility of building a WHR unit in these plants was evaluated.

# 1. INTRODUCTION

The production of cement is an intensive resource that including large amounts of raw materials, labor, energy, and capital. Chalk, sand, limestone, shale, and clay are the main raw composed materials of cement. Moreover, the low value of other materials like alumina, iron ore, and other minerals are added to regulate the composition of raw material [1-5]. These raw materials are quarried, crushed, finely ground, and blended to the appropriate chemical composition. The raw materials are usually heated to 1724 K in a rotary kiln to form a clinker. Clinker is cooled and ground with other minor additives and gypsum to produce cement [3-9].

The feed rate of the rotary kiln is regulated by the speed of the rotational and the slope of the kiln. Natural gas, coal and pet coke more progressively, alternative fuels like solvents, plastic, waste oil, and bone meal are fed to the kiln and being on fire to feed the flame, which can reach temperature of 2000 K [4-10]. As the kiln very slow rotates, the raw materials fall over progressively hot zones near the flame at the end of the kiln. Inside the kiln's burning zone, raw materials reach temperatures of 1700-1900 K. At about 1750 K, a set of chemical reactions leads the materials to decompose, in a molten state, and fuse to nodules [5, 6]. Hot exit gases going through the kiln are used to preheat and calcine the raw materials feed, which they come from the kiln's burning zone. Clinker is discharged from the end of the kiln to air coolers to handling temperatures. Cooled clinker is mixed with additives and gypsum and ground in a gray powder called cement. Many cement plants contain the final cement grinding and mixing operation at the site [6, 7].

Rotary kilns are based on produced raw materials. In wet-process kilns, raw materials are fed to the kiln, such as slurry with 30-40% moisture. The wet process needs higher energy because of the water amount that must be evaporated before calcination [1, 11-14]. Generally, a wet-process kiln needs an additional length and nearly 100% more kiln thermal energy compared to an efficient dry kiln. Three types of dry-process kilns are the long dry kilns without any preheater, suspension preheater kilns and pre-calciner or new suspension preheater kilns [2-4, 8]. In suspension preheater and new suspension preheater kilns, the pyro-processing initial sections happen in the preheater stages before materials come into the rotary kiln. As the raw materials are passed down over cyclones, they are in contact with hot kiln exhaust gases in a reverse direction; consequently, heat is transferred from the gas to the materials. This preheats and partially calcines the materials before they enter the kiln, which leads the main chemical reaction stake place more quickly and efficiently [7-10]. Depending on the moisture content of the raw materials, a kiln may have three to six stages of cyclones with increasing heat recovery with each extra stage. Therefore, suspension preheater and new suspension preheater kilns have higher production capacities and greater fuel efficiency compared to other types of cement kilns [5, 7, 15]. Table 1 compares a typical thermal energy consumption by wet and dry rotary kiln types.

During pyro-processing, three central processes happen with a mixture of raw materials. First, all moisture is driven off; after that, the calcium carbonate in limestone dissociates to carbon dioxide and calcium oxide in a process called calcination. Finally, the lime and other minerals in the raw materials react to form calcium silicates and calcium aluminates, which are the main components of clinker in a process known as sintering [1, 3, 9-16]. After the clinker is shaped in the rotary kiln, it is cooled rapidly to minimize glass phase generation and confirm the maximum yield of alite formation, a central component for cement hardening properties. In the grate cooler, the clinker is transported over a reciprocating grate through which air flows perpendicular to the clinker flow. In the planetary cooler, the clinker is cooled with air and in a counter-current stream. The cooling air is utilized as secondary combustion air for the kiln [3, 17]. Clinker nodules are ground to the consistency of powder in a ball mill. Clinker grinding, composed with additions of approximately 5% gypsum to control setting properties of cement, can be ready in ball mills combined with roller presses, vertical roller presses, or roller mills [16-21].

In this chapter, the most effective technologies based on the Rankine cycle, which is applied for recovering waste heat in the cement sector are explained. Besides, the economic feasibility of waste heat recovery in three cement plants around the world is studied.

**Table 1. Consumption of specific thermal energy using rotary kiln [1]**

| Kiln type | Heat input (kMBtu/tonne of clinker) | Heat input (kJ/tonne of clinker) |
|---|---|---|
| Long Dry (LD) | 4360 | 4600000 |
| Wet | 5550-5950 | 5860000-6280000 |
| 1 Stage Cyclone Preheater (SP) | 3960 | 4180000 |
| 2 Stage Cyclone Preheater (SP) | 3570 | 3770000 |
| 4 Stage Cyclone Preheater (SP) | 3360 | 3550000 |
| 4 Stage Cyclone Preheater plus Calciner (new suspension preheater) | 2970 | 3140000 |
| 5 Stage Cyclone Preheater plus Calciner (new suspension preheater) plus high efficiency cooler | 2850 | 3010000 |
| 6 Stage Cyclone Preheater plus Calciner (new suspension preheater) plus high efficiency cooler | 2780 | < 2930000 |

## 2. Waste Heat Recovery in the Cement Process

New suspension process kilns contain multi-stage preheaters to preprocess raw materials before they arrive at the kiln and an air-quench system to cool the clinker. Kiln exhaust streams, from the clinker cooler and the kiln preheater system, include beneficial thermal energy that can be transformed into power generation [1, 22-24]. Maximizing the total efficiency of the kiln process is paramount for the effective operation of the plant. However, residual waste heat from the clinker coolers and preheater exhausts can be improved and used to afford low-temperature heating requires in the plant, or used to generate power to balance a portion of power purchased from the grid, or captive power created by fuel consumption at the site [25]. Usually, cement plants do not have substantial low-temperature heating requests; therefore, most waste heat recovery projects have been for power generation. The volume of waste heat existing for recovery depends on the production and design of the kiln system, the raw materials moisture content, and the quantity of heat needed for drying in the raw mill system, stable fuel system, and cement mill. Waste heat recovery can produce up to 30% of the consumed cement plant's overall electricity and has the following benefits [1, 24-28]:

- Decreases purchased power consumption, which in turn decreases the costs of operating.
- Reduces the effect of increasing electricity price in the future.
- Increases reliability of plant power.
- Raises plant competitive position in the market.
- Decreasing greenhouse gas releases.

## 2.1. Waste Heat Recovery Power Systems

Waste heat recovery power systems applied for cement kilns are run based on the Rankine Cycle. This thermodynamic cycle is the base for current thermal power-producing stations. It includes a heat source (boiler) that changes a liquid to high-pressure vapor, which is then expanded over a turbo-generator creating power [29-33]. Low-pressure vapor exhausted of the turbo-generator is condensed back to a liquid state, with condensate from the condenser given back to the boiler feed-water pump to remain the cycle. Waste heat recovery systems consist of heat exchangers that transport heat from the exhaust gases to the working fluid, turbines, electric generators, condensers, and a working fluid cooling system. Three primary waste heat recovery power generation systems are accessible, distinguished based on the type of working fluid as follows [1, 26-35].

### *2.1.1. Steam Rankine Cycle*

The most usually used Rankine cycle system for waste heat recovery power generation utilizes water as working fluid, and includes generating steam in a waste heat boiler, which then drives a steam turbine. Steam turbines are one of the first and most useful power generation technologies in use [1]. In the waste heat recovery steam cycle, the working fluid is at the first step pumped to elevated pressure before entering the waste heat recovery boiler. The water is vaporized as high-pressure steam using the hot exhaust from the process and after that, expanded to lower temperature and pressure in a turbine, generating mechanical power that drives an electric generator [36, 37]. The low-pressure steam is after that exhausted to a condenser under vacuum conditions, where the expanded vapor is condensed to low-pressure liquid and refunded to the feed-water pump and boiler. In cement plants, steam cycles are used for the most communal waste heat recovery systems in operation and have the following advantages and disadvantages [34-40]:

- Most well-known to the cement industry and are usually economically better where source heat temperature is higher than about 575 K.
- Based on proven technologies and usually naive to operate.
- Extensively available from a change of suppliers.
- Commonly have minimum installation costs than Rankine cycle systems.
- Need higher temperature waste heat to operate optimally.
- Need a full-time operator, related to local regulations.
- Need feed-water conditioning systems.
- Need a cooled water condenser; air-cooled condensers can be utilized but generate a performance penalty because of higher condenser vacuum pressures.
- Generally, match well with huge kilns and systems with low raw material water value.

*2.1.2. Organic Rankine Cycle*

Other types of working fluids with well production efficiencies at lower heat source temperatures are utilized in an organic Rankine cycle (ORC) system [1]. The ORCs usually utilize an organic as working fluid like pentane or butane with a lower boiling point, higher molecular mass, higher vapor pressure, and higher mass flow rate relative to water. Together, these properties allow higher turbine efficiencies than those accessible by a steam system [24, 41]. The ORC systems can be used for waste heat sources as low as about 420 K, but steam systems are restricted to heat sources more than about 530 K. The ORC systems are usually planned with two heat transfer sections. The first part transfers heat from the waste gases to an intermediate heat transfer fluid. The second part transfers heat from the middle heat transfer fluid to the organic working fluid [33-36]. The ORCs have usually been utilized to generate power in geothermal power plants, and more newly, to recover heat from pipeline compressors in the United States. The ORC systems have been broadly utilized to generate power from biomass systems in Europe. A few ORC systems have been installed on cement kilns. The ORC's specific properties contain the following [40-48]:

- It can improve heat from gases at lower temperatures than is promising with usual steam systems, the ability of ORCs to use all recoverable heat from the air cooler.
- Operate with condensing systems above the pressure of atmospheric, decreasing risk of air leakage to the system and removing the necessity for a de-aerator.
- Not sensitive to freezing.
- Due to ORCs operate at rather low pressure; they can operate unattended and wholly automatically in many locations related to local regulations.
- The organic fluid features result in the working fluid remaining dry (without any partial condensation) during the turbine, evading blade erosion.
- Can use air-cooled condensers without upside down impacting performance.
- Lower speed ORC turbine permits generator direct drive without the required for and inefficiency of a decrease gear.
- ORC apparatus (piping, condensers, turbines, heat exchanger surface) is usually smaller than that needed for steam systems, and the turbine usually composed of fewer stages.

- Though ORCs can prepare generated efficiencies equal to a steam Rankine system, ORCs are usually applied to lower temperature exhaust streams and restricted in sizing and scalability and usually are smaller in capacity than steam systems.
- Referring to the application, ORC systems usually have a higher particular cost than steam systems.
- The two-stage heat transfer process forms some system inefficiencies.
- The heat transfer fluids and organic fluids usually utilized in ORCs are flammable, needing fire safety measures, and occasional additional overtime. Furthermore, there may be environmental concerns over potential system leaks.
- In total, ORC systems are better matched with excellent efficiency, small to medium size kilns or kilns with elevated raw material moisture content.

### *2.1.3. Kalina Cycle*

The Kalina Cycle is another Rankine cycle that utilizes a binary mixture of water and ammonia as the working fluid, which permits for more effective energy extraction from the heat source. The Kalina cycle takes advantage of the ability of ammonia-water mixtures to use changing and lower temperature heat sources [1]. The Kalina cycle has an operating temperature range that can agree with waste heat at temperatures of 370-810 K and is declared to be 15-25% more effective than ORCs at a similar temperature level. The Kalina cycle is in the market summary, with a total of nine operating systems in various industries like refining and steel and in geothermal power plants where the hot fluid is very often a liquid below about 420 K [36, 39]. Kalina cycle systems are now being piloted in the cement industry. The key characteristics of the Kalina cycle are as follow [47-51]:

- It can be utilized in lower temperature applications than usual steam Rankine cycle systems.
- Very flexible; the system has a high turndown ratio and fast answer to variations in flow and heat source temperature.

- The ammonia-water binary mixture can be proctored to attain improved heat transfer and greater efficiency using similar flows and waste heat temperatures.
- The binary mixture as the working fluid is nonflammable.
- The technology is in the primary stage of market introduction with restricted experience and suppliers.

## 2.2. Application of Waste Heat Recovery Power Systems in the Cement Process

Japanese companies have headed the overview of steam cycle waste heat recovery power systems in the cement industry [1, 39]. In 1980, Kawasaki Heavy Industries performed the first waste heat recovery system in operation at Sumitomo Osaka Cement. The first primary usual system, with a capacity of 15 MW, has been in setup since 1982 at Taiheiyo Cement's Kumagaya plant. China established the first system in 1998 in association with a Japanese supplier. Clean development mechanism and Government policy motivations began to drive the market in China, and over 700 units were operating in China in 2012 [1]. The bulk of market acting today is in Asia; the companies of Chinese or shared ventures are the initial suppliers. The waste heat recovery systems leading manufacturers by usual steam path technology are now marketing second production systems with greater parameters of supercritical steam and reclaimed efficiencies that reach output levels as high as 45 kWh/t of clinker [1, 44].

In a usual waste heat recovery system erection on a new suspension preheater kiln, waste heat boilers are erected on the hot exhaust streams leaving the preheaters and air quells clinker cooler to generate medium or low-pressure steam [49]. The steam is fed to a condensing steam turbine that drives a generator to create power. Hot condensate of the condenser is fed back to the waste heat boilers. The whole system consists of the air quench clinker cooler waste heat boilers, the steam turbine generator, and auxiliary apparatus like water treatment system, condenser, recoiling system, and boiler feed pump [1].

Referring to the number of preheater stages (2-6 preheater stages), the exhaust temperatures from a new suspension preheater kiln system usually are in the range of 555-725 K, while the temperatures of waste air from the clinker cooler are usually in the range of 524-604 K. For example, in the case of a $3\times10^3$ tpd (tonne/day) production line of clinker, about $1.7\times10^5$ $Nm^3/h$ of kiln exhausts and $1.5\times10^5$ $Nm^3/h$ of cooler air is created [1]. Usage of waste heat recovery power systems in the cement process kilns can be challenged. The exhaust gases from the clinker cooler and the kiln preheaters usually consist of rather high concentrations of dust that periodically be over 50 $g/N^3m$, and the temperatures of waste gas can undulate broadly through kiln operation. Furthermore, many plants use the exhaust gas to raw materials drying, and the amount available for heat recovery can vary broadly, referring to the content of raw feed moisture [41].

## 2.3. Recoverable Waste Heat and the Potential for Power Generation

The value of recoverable waste heat from a new suspension preheater kiln referring to various factors containing the following [1, 40, 46, 50]:

- Raw material feed moisture content.
- Value of additional air in the kiln.
- Value of air infiltration.
- Efficiency and number of preheater and pre-calciner sections.
- Formation of the clinker cooler system.

The numbers of preheater sections in a cement plant have a substantial bearing on the potential of waste heat recovery and the consumption of total thermal energy. The higher the number of stages, the higher the total efficiency of thermal energy of the kiln and the lower the potential for waste heat recovery. The selection of the number of preheater stages is based on various factors like restrictions on preheater tower height, cooler efficiency, and requirements of heat for the mill itself [51]. Table 2 gives the waste heat recoverable quantity from new suspension preheater kilns. Temperatures of preheater exhaust are in the range of about 665 K for small kilns with four preheater sections, to under about 575 K for large kilns with six preheater stages.

Figure 1 indicates the power generation potential for a steam waste heat recovery system applied exhaust to a typical 5000 tpd clinker line for exhaust temperatures ranging from about 575-725 K. Furthermore, the design of the clinker cooler impacts wastes heat accessibility [1]. The fundamental cooler function is to eliminate heat from hot clinker discharged from the kiln. Moreover, fast cooling makes better grind-ability and the quality of clinker. Usually, new coolers are grate, which have several stages of development [46-49]. Table 3 provides the heat available in several grate cooler generations. Air temperatures of exhaust from the clinker cooler are in the range of about 525-605 K, referring to the configuration of cooler and efficiency of recuperation.

**Table 2. Typical available heat for dry process new suspension preheater kilns [1]**

| Parameter | Unit | Preheater kilns | Number of preheater with precalciner stages | Number of preheater with precalciner stages | Number of preheater with precalciner stages |
|---|---|---|---|---|---|
| Number of cyclone stages | - | 4 | 4 | 5 | 6 |
| Top stage exit temperature | Deg K | 664 | 634 | 590 | 556 |
| Kiln capacity range | TPD | 1000-2500 | 2000-8000 | 2000-8000 | 2000-8000 |
| Heat available in preheater exhaust | kJ/hr for 1 MTPA* (Mkcal/hr) | 113000 (27) | 94300 (22.5) | 81100 (19.4) | 73300 (17.5) |
| Specific heat consumption | kJ/ tonne clinker (cal/g) | 3550 (850) | 3140 (750) | 3010 (720) | 2930 (700) |
| Heat available in preheater exhaust | kJ /tonne clinker (cal/g) | 904 (216) | 754 (180) | 649 (155) | 586 (140) |

*MTPA Million metric tonnes per annum.

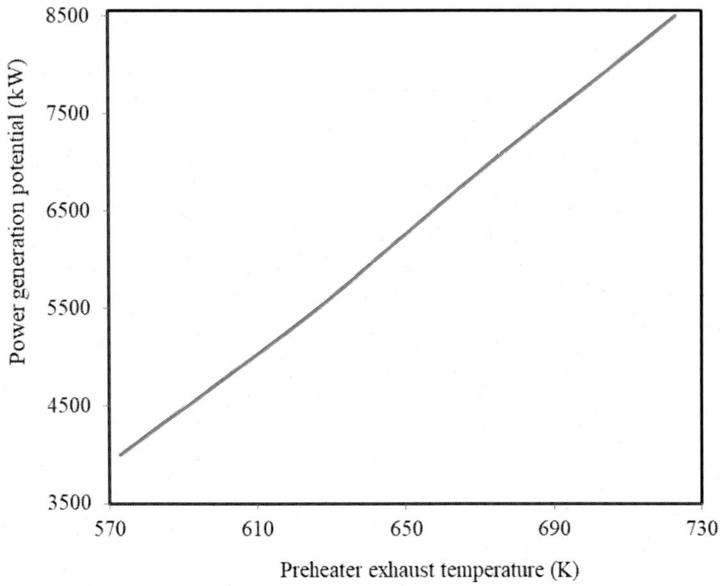


Figure 1. Variations of power generation potential [1].

**Table 3. Typical available heat for grate clinker coolers [1]**

| Parameter | Unit | 1st Generation | 2st Generation | 3st Generation |
|---|---|---|---|---|
| Grate plate type | - | Vertical aeration with holes in plate | Horizontal aeration | Horizontal aeration |
| Heat available in exhaust | kJ/ tonne clinker (cal/g) | 419-505 (100-120) | 335-419 (80-100) | 293-335 (70-80) |
| Cooling air input | Nm3/kg clinker | 2-2.5 | 1.8-2 | 1.4-1.5 |
| Exhaust air volume | Nm3/kg clinker | 11.5 | 0.9-1.2 | 0.7-0.9 |
| Heat available in exhaust | kJ/hr for 1 MTPA (Mkcal/hr) | 52300-62800 (12.5-15) | 41900-52300 (10-12.5) | 36600-41900 (8.8-10) |
| Recuperation efficiency | % | < 65 | < 70 | > 73 |

Table 4 gives the total heat available from both the preheater exhaust and clinker cooler air for a typical $5\times10^3$ tpd clinker plant. Power conversion efficiencies in the range of 18-25% result in potential power capacities of 6-9 $\times10^6$ W [1].

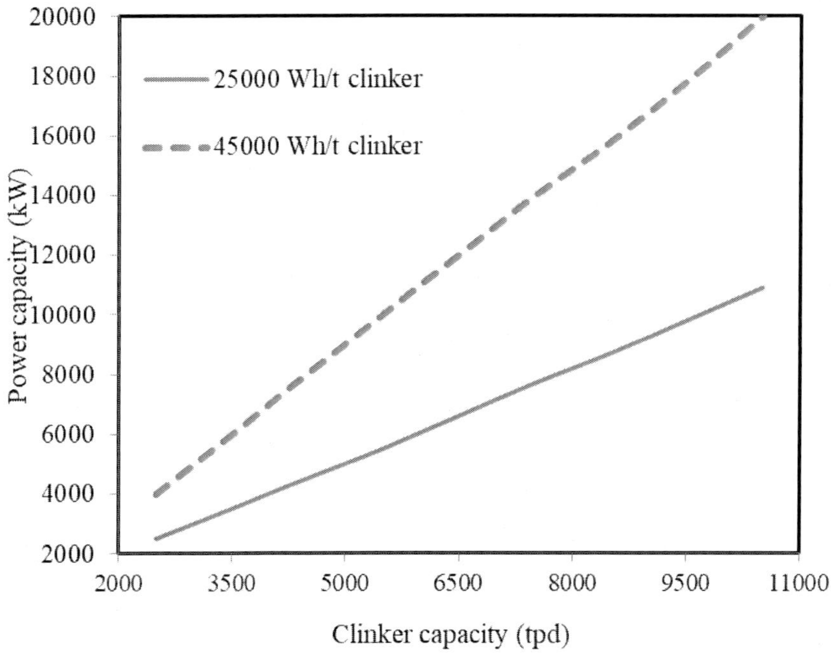


Figure 2. Waste heat power generation capacities [1].

Assuming an average plant electrical drive power requisite of $1.06\times10^5$ Wh/t of cement and a clinker factor of 0.75, around 20-30% of the needed electricity for the production of cement process can be made from the waste heat [1]. Figure 2 indicates the band of expected power generation for a range of kiln capacities. An extra limiting factor in the heat available for effective recovery is the content of moisture of the raw material incoming the kiln.

The moisture content of limestone deposits can range from 2-15%, referring to the origin of limestone. The quantity of moisture presents in the feed material incoming the kiln preheater affects the particular consumption of heat in the kiln and the rate of kiln production. The usual practice is to restrict moisture content incoming the kiln to less than 1%. Therefore, the materials of raw feed are typically dried during grinding using cooler hot gas and preheater hot gas as the heat source [25, 30, 44].

In order to show the moisture impact on drying needs, the raw material flows and heat needed in Mkcal/h for several kiln production rates at various moisture levels referring to the following assumptions are given in Table 4:

- Raw mill to clinker factor of 1.55.
- Heat requirement of 3.98 GJ/ton of water for the raw mill (950 kcal/kg).
- Raw mill running hours per day.

**Table 4. Typical available heat and power generation from preheater/grate clinker cooler [1]**

| 5000 tpd clinker line, 100% utilization of available waste heat | |
|---|---|
| Input heat to preheater waste heat and quench clinker cooler boilers | 963 kJ/tonne clinker (230 cal/g) |
| Heat available for power (input -output) | 583 kJ/tonne clinker (1400 cal/g) |
| Output heat in boiler exhaust gas | 379 kJ/tonne clinker (90 cal/g) |
| Potential power generation | 6000-9000 kW |
| Power conversion efficiency | 18-25% |

As observed in Table 4, a significant amount of heat is needed to dry raw martial in the high moisture levels. High moisture amount in the raw feed can notably decrease the heat available for waste heat recovery in the preheater exhaust; raw feeds with very high moisture rates fundamentally limit the effective heat recovery potential. Usually, the content of raw feed moisture lower than 4-6% would have minimal impact on waste heat recovery potential. At higher moisture levels, waste heat recovery viability depends on operating conditions, size of kiln, and properties of raw material [1].

**Table 5. Heat required for raw material drying [1]**

| Raw Material Flow, TPD | 3382 | 5073 | 6774 | 8455 | 10145 | 11836 | 13527 |
|---|---|---|---|---|---|---|---|
| Kiln Capacity, TPD | 2000 | 3000 | 4000 | 5000 | 6000 | 7000 | 8000 |
| Raw Moisture Content | | | | Drying heat required, GJ/min | | | |
| 2 percent | 0.12 | 0.17 | 0.23 | 0.28 | 0.34 | 0.39 | 0.46 |
| 4 percent | 0.39 | 0.59 | 0.79 | 0.97 | 1.17 | 1.36 | 1.56 |
| 6 percent | 0.55 | 0.82 | 1.09 | 1.36 | 1.63 | 1.89 | 2.17 |
| 8 percent | 0.81 | 1.23 | 1.63 | 2.03 | 2.44 | 2.84 | 3.26 |
| 10 percent | 1 | 1.53 | 2.04 | 2.55 | 3.06 | 3.57 | 4.29 |
| 12 percent | 1.28 | 1.92 | 2.55 | 3.19 | 3.83 | 4.46 | 5.09 |

As discussed previous, heat can be recovered from both the clinker cooler exhausts and preheater. There are three types of steam cycle systems, and each has advantages and disadvantages that are particular to installation needs, the relative heat available quantities, and the temperatures of the exhaust of the coolers and preheaters [1, 42, 50].

### *2.3.1. Single-Pressure System*

This is the most straightforward system and is based on utilizing the equal steam pressure in both the Pre-heater and quench clinker cooler boilers. Usually, feed-water from the boiler pump is preheated using the economizer section of the quench clinker cooler boiler and after that split into two flows. One flow goes out to the quench clinker cooler boiler drum and the other to the PH boiler drum [1]. Superheated steam from both boilers (usually11.5 bar, 585-615 K) is incorporated in a single header and forwarded to the turbine generator. The temperatures of the exhaust are decreased to between about 365-385 K with a single pressure system. System design is simple and usually lower cost; however, fluctuations in temperature of exhaust gas and flows are not used well, leading to inefficiencies in heat recovery subject to off-design conditions [50, 51].

### *2.3.2. Flash Evaporator System*

This system varied from the single-pressure system principally in which the feed-water left from the air quench clinker cooler economizer is split into three flows. The first two flows are equal to the one-pressure system (one comes to the PH boiler and another one to the air quench clinker cooler boiler) at the same pressure. The third flow is forwarded to a flash evaporator to produce low-pressure (1-2 bar) saturated steam. Then, this low pressure steam is forwarded to the midpoint in a dual-pressure steam turbine. Feed-water that is not flashed to steam is returned as boiler feed-water. The flash evaporator lets extra system flexibility for oscillations in the hot gas streams temperatures or flows. Exit temperatures of the exhaust gases can be controlled at about 365 K, avouch that resources of waste heat are wholly utilized. The water content of the steam incoming the turbine midpoint is higher; subsequently, the last stage turbine blade is subject to raised erosion [1, 43, 47].

### *2.3.3. Dual-Pressure System*

This system comprises two drums at various pressures in the air quench clinker cooler boiler. The feed-water left from the economizer is divided into three parts. The first part is sent to the high-pressure air quench clinker cooler drum while the second part to the PH boiler (at the same pressure). The third part is sent to the low-pressure air quench clinker cooler boiler drum, still producing superheated steam. This low-pressure steam is forwarded to the midpoint of the dual-pressure steam turbine. Similar to the flash evaporation system, this leads system flexibility for fluctuations in temperatures of exhaust and flows; however, the superheat steam decreased the last stage blades erosion in the steam turbine [1].

## 3. Process Description of the Plant Production Line

In a cement plant, clinker is produced in the pyro-processing system that consists of a rotary kiln and a pre-heater section. The diameter of rotary kilns is generally between 3-6 m, and the rotational speed of them is between 2–3 rpm. Cyclone type pre-heaters are vastly used in the new generation pyro-processing system to pre-heat raw materials before entering the rotary kiln. The pre-calcination becomes to start inside the pre-heater, and nearly 95% of kiln feed might be pre-calcined at the kiln intake. The kiln feed passes across the rotary kiln, and the clinker's phases, like alite and belite are formed in the burning zone. Then, clinkers fall in a grate cooler to improve the product and recover heat from clinker. In this chapter, the possibility of electricity generation using WHR technology is studied for three Portland cement plants in Nigeria (plant 1), Turkey (plant 2), and Iran (plant 3). The energy analyses of these plants were performed in previous works [35, 42, 52]. The operation conditions of the pyro-processing system of them are illustrated in Figures 3-5.

## 4. Waste Heat Recovery

Nowadays, installing a WHR unit in a cement plant in order to recover waste heat is an economical method to produce electricity. The main portion of lost in a cement factory is the thermal energy of pyro-processing exhaust gases, which would be recovered in a WHR unit [36]. A part of the thermal energy of the exhaust gases could be utilized to drive a turbine in the WHR unit to generate electricity. The WHR unit could generate a noticeable part of electricity to consume in a cement plant. To determine the scale of the

WHR unit, the available heat energy from the exhaust gases should be specified.

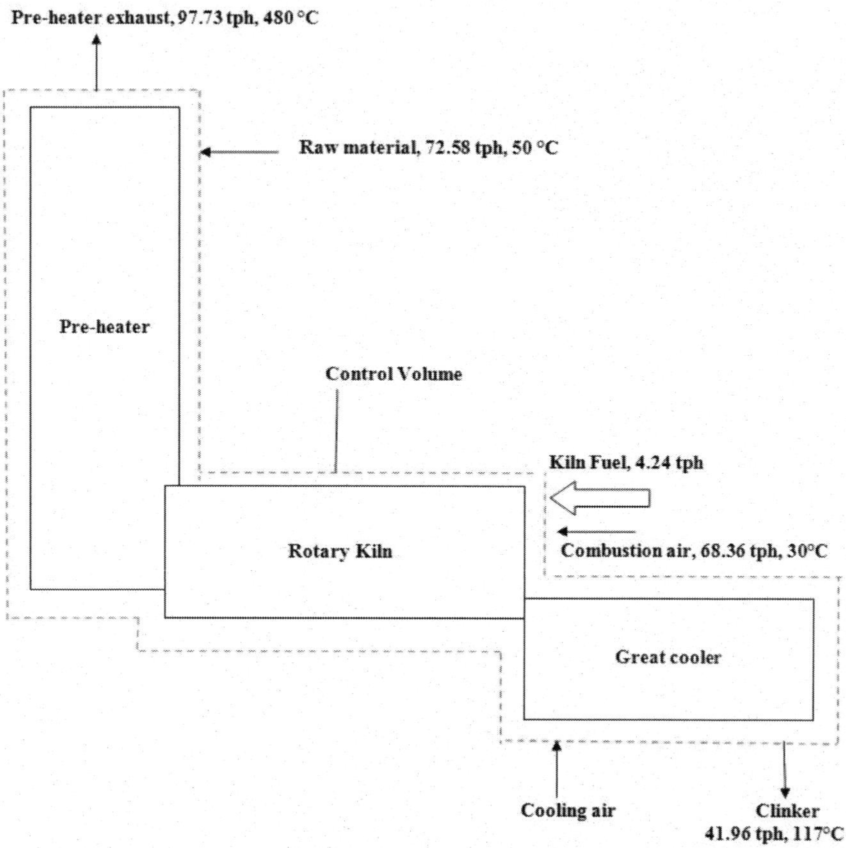


Figure 3. Flow diagram of pyro-processing system for cement plant 1 (Nigeria) [52].

## 4.1. Thermal Energy

Regarding the energy audits, which were performed in the pyro-processing systems [35, 42, 52], the total quantities of thermal energy that were lost from stacks in plants 1, 2 and 3 were 1217, 706 and 672 $kJ/kg_{cl}$, respectively. A part of these thermal energies must be utilized to dehumidify the mill feed in order to increase the efficiency of the roller or ball mill, which is used to grind raw materials [53]. Generally, the installation of a waste heat recovery unit will not be economical if the raw materials' moisture is greater than 6 wt% since the main part of the thermal energy of the exhaust gases should be used for drying mill's feed. In addition, the minimum temperature of output gas from the waste heat recovery unit is usually about 100 °C because, at a lower temperature, the efficiency of this unit decreases dramatically.

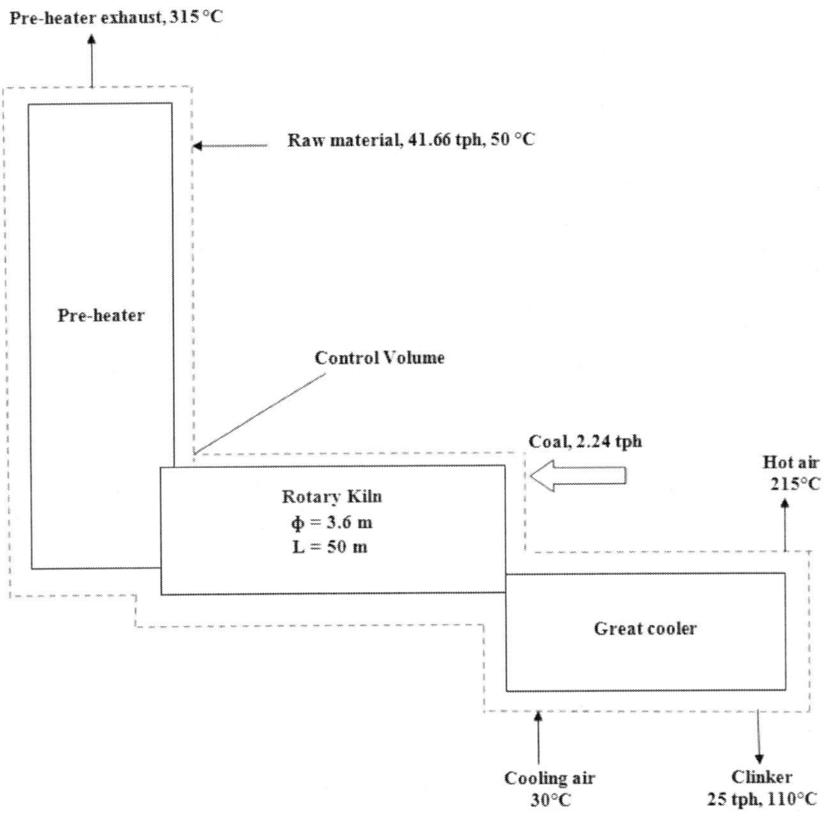


Figure 4. Flow diagram of pyro-processing system for cement plant 2 (Turkey) [35].

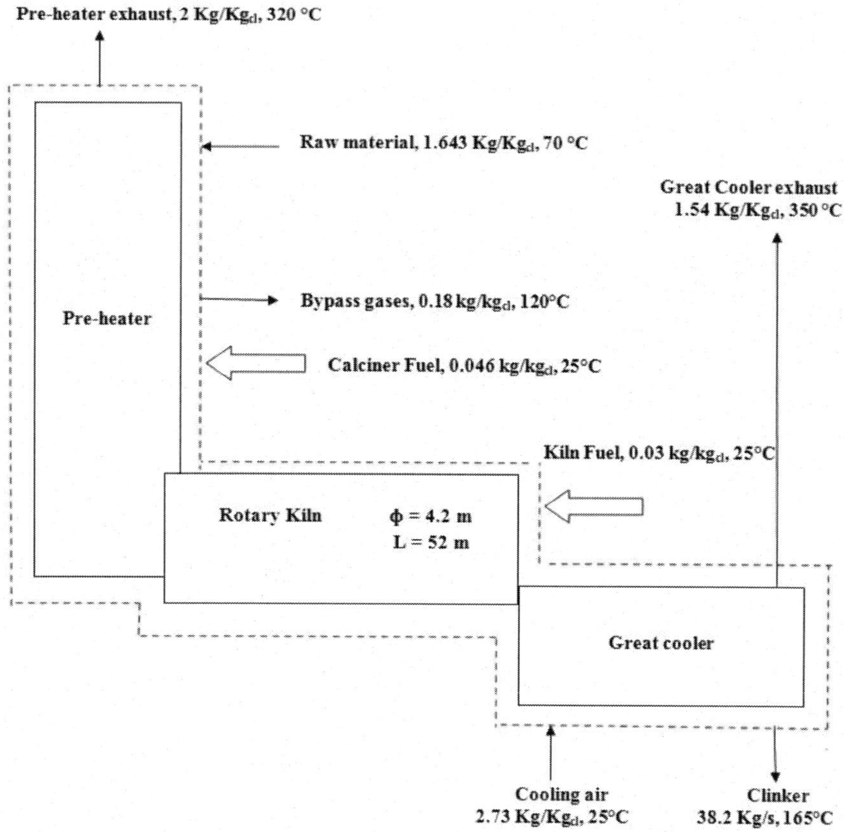


Figure 5. Flow diagram of pyro-processing system for cement plant 3 (Iran) [42].

The portion of lost thermal energy that would be used to produce electricity is calculated using the following equation:

$$Q = \dot{M}_g C_{Pg} (\theta_2 - \theta_1) \tag{1}$$

Where, $\dot{M}_g$ is mass flow rate (kg/s), $C_{Pg}$ is specific heat capacity (kJ/kg. °C), and θ is gas temperature (°C).

It can be seen in Table 6 that the total heat energies, which are available to generate electricity by using the WHR unit, are 811, 458 and 744 kJ/$kg_{cl}$ for plants 1, 2 and 3, respectively.

**Table 6. The available heat energy of the pyro-processing systems for the WHR unit**

| | Plant 1 (kJ/kg) | Plant 2 (kJ/kg) | Plant 3 (kJ/kg) |
|---|---|---|---|
| The heat energy of the exhaust gases | 1217 | 706 | 672 |
| The available heat energy of pre-heater exhaust gases for producing electricity | 811 | 347 | 336 |
| The heat energy of grate-cooler exhaust gases | - | 215 | 572 |
| The available heat energy of grate-cooler exhaust gases for producing electricity | - | 111 | 408 |
| The total available heat energy for producing electricity | 811 | 458 | 744 |

## 4.2. Electrical Energy

The potential of electricity generation from the waste heat in a cement plant depends on the temperature of exhaust gases from stacks, the mass flow of exhaust gases from stacks, the moisture content of the raw mill's feed, and the efficiency of the WHR unit.

**Table 7. The potential of electricity generation in plant 1 from heat recovery of the exhaust gases using a WHR unit**

| ORC | SRC |
|---|---|
| WHR efficiency 25% | 23.5% |
| Produced electrical energy 0.0561 | 0.0528 kW/kgcl |
| Clinker production 41960 kg/h | 41960 kg/h |
| Total produced electrical energy 2.35 MWh | 2.22 MWh |
| Electrical energy that is used in the WHR (7%) 0.16 MWh | 0.16 MWh |
| Net produced electrical energy 2.19 MWh | 2.06 MWh |

As seen in Tables 7-9, the potential of net electric power generation using an ORC unit with energy conversion efficiency accounts for 25% from the waste heat of plants 1, 2 and 3 are 56.1, 37.8 and 51.5 kW/t, respectively. Besides, the potential of net electric power production using an SRC unit with energy conversion efficiency accounts for 23.5% from the waste heat of plants 1, 2 and 3 are 52.8, 29.8 and 43.2 kW/t, respectively. The potential of net electric power generation by waste heat recovery for these cement plants is tabulated in Tables 7-9.

**Table 8. The potential of electricity generation in plant 2 from heat recovery of the exhaust gases using a WHR unit**

| ORC | SRC |
|---|---|
| WHR efficiency 25% | 23.5% |
| Produced electrical energy 0.0378 kW/kgcl | 0.0298 kW/kgcl |
| Clinker production 25000 kg/h | 25000 kg/h |
| Total produced electrical energy 0.945 MWh | 0.745 MWh |
| Electrical energy that is used in the WHR (7%) 0.052 MWh | 0.052 MWh |
| Net produced electrical energy 0.893 MWh | 0.693 MWh |

**Table 9. The potential of electricity generation in plant 3 from heat recovery of the exhaust gases using a WHR unit**

| ORC | SRC |
|---|---|
| WHR efficiency 25% | 23.5% |
| Produced electrical energy 0.0515 kW/kgcl | 0.0432 kW/kgcl |
| Clinker production 137,500 kg/h | 137,500 kg/h |
| Total produced electrical energy 7.08 MWh | 5.94 MWh |
| Electrical energy that is used in the WHR (7%) 0.5 MWh | 0.42 MWh |
| Net produced electrical energy 6.58 MWh | 5.52 MWh |

## 4.3. Economic Feasibility

The primary capital to establish a WHR unit includes the land, design of the system, equipment, construction structures, and engineering services. The initial investment cost would change based on the type of WHR technology, the factory's location and the main lines of electricity distribution. In addition, the needful cost of installing the WHR unit depends on the capacity, and dimension of the WHR unit since by decreasing the unit dimensions, the initial investment cost is increased.

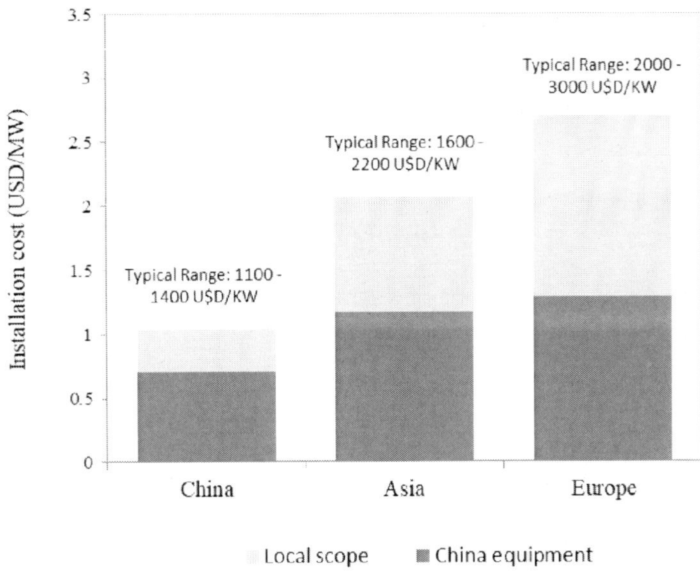


Figure 6. The price of Chinese WHR units [42].

Nowadays, the initial investment of the WHR unit in China is less than that of in the other countries. Figure 6 illustrates the amount of initial investment of Chinese WHR units in some regions of the world. The price of Chinese WHR units is usually approximately 20-30% lower than that of similar European WHR units [54, 39].

Some factors could affect the capital return rate that are as follow [55, 56]:

- The price of WHR units
- The expense of maintenance and operation of the unit, which is approximately 2.5% of the price of the WHR unit per annum.
- The operating period of the Pyro-processing system (hours per annum)
- Electricity price (USD/MW)
- The quantity of net electricity production per annum

The details of the economic feasibility of installing the WHR unit for the plants are provided in Tables 10-12. As can be seen in these tables, the capital return rates of plant 3 are lower than those of the other plants because the electricity price is very low (0.03 USD/kW) in Iran, and the capital return rate of plant 2 is more than that of plant 1 because of higher electricity price.

**Table 10. The economic feasibility for a WHR unit establishment in the plant 1**

| WHR payback calculations ORC | SRC |
|---|---|
| Clinker manufactured capacity 41960 kg/h | 41960 kg/h |
| Net produced electrical energy 2.19 MWh | 2.06 MWh |
| Annual kiln working hours 7200 h | 7200 h |
| Annual net produced electrical energy 15678 MWh | 14832MWh |
| The electrical energy price 0.07 USD/kW | 0.07 USD/kW |
| Annual gross income 1097460 USD | 1038240 USD |
| Annual maintenance cost (2.5%) 27437 USD | 25956 USD |
| Tax (30%) 329238 USD | 311472 USD |
| Annual primary capital 740785 USD | 700812 USD |
| Total primary capital 3520000-4840000 USD<br>The capital return rate (in the year)21-15.3% | 3360000-4620000 USD<br>20.8-15.1% |
| The capital return time 4.76-6.53 years | 4.79-6.59 years |

**Table 11. The economic feasibility for a WHR unit establishment in the plant 2**

| WHR payback calculations ORC | SRC |
|---|---|
| Clinker manufactured capacity 25000 kg/h | 25000 kg/h |
| Net produced electrical energy 0.893 MWh | 0.693 MWh |
| Annual kiln working hours 7200 h | 7200 h |
| Annual net produced electrical energy 6430 MWh | 4990 MWh |
| The electrical energy price 0.09 USD/kW | 0.09 USD/kW |
| Annual gross income 578700 USD | 449100 USD |
| Annual maintenance cost (2.5%) 14468 USD | 11228 USD |
| Tax (20%) 115740 USD | 89820 USD |
| Annual primary capital 448492 USD | 348052 USD |
| Total primary capital 1440000-1980000 USD<br>The capital return rate (in the year) 31.1- 22.6% | 120000-1540000 USD<br>31- 22.6% |
| The capital return time 3.21- 4.42 years | 3.22-4.42- years |

**Table 12. The economic feasibility for a WHR unit establishment in the plant 3**

| WHRPG payback calculations ORC | SRC |
|---|---|
| Clinker manufactured capacity 137500 kg/h | 137500 kg/h |
| Net produced electrical energy 6.58 MWh | 5.52 MWh |
| Annual kiln working hours 7200 h | 7200 h |
| Annual net produced electrical energy 47376 MWh | 39744 MWh |
| The electrical energy price 0.03 USD/kW | 0.03 USD/kW |
| Annual gross income 1421280 USD | 1192320 USD |
| Annual maintenance cost (2.5%) 35532 USD | 29808 USD |
| Tax (25%) 355320 USD | 327888 USD |
| Annual primary capital 1030428 USD | 864432 USD |
| Total primary capital 10560000-14520000 USD<br>The capital return rate (in the year) 10.3-7.1% | 8960000-12320000 USD<br>9.6-7% |
| The capital return time 10.22- 14.05 years | 10.3-14.1 years |

Figures 7 and 8 show the capital return time of plants 1, 2 and 3 based on different initial investments of the WHR unit. For both SRC and ORC technologies, this rate for plant 2 is lower than for plants 1 and 3; therefore, it seems that installing the WHR unit for plant 2 is more economical than plants 1 and 3.

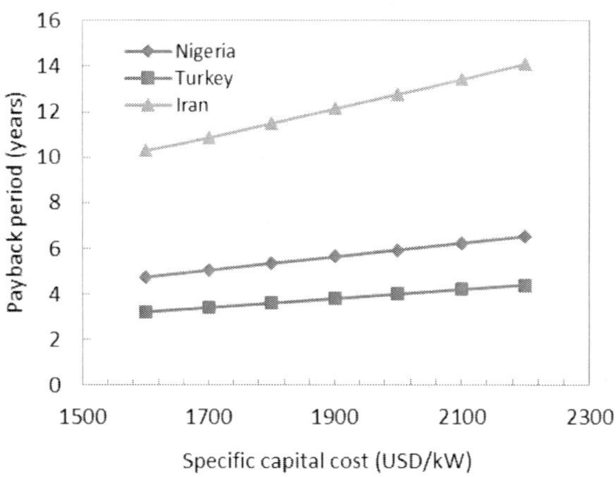


Figure 7. The payback period of the investment for different primary capital costs of ORC technology.

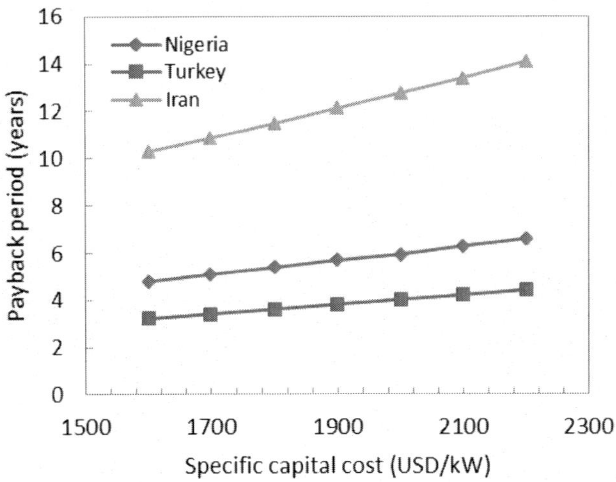


Figure 8. The payback period of the investment for different primary capital costs for SRC technology.

## CONCLUSION

In this chapter, first, three types of waste heat recovery technologies, which are named steam ranking cycle (SRC), organic ranking cycle (ORC), and Kalina ranking cycle, were discussed for utilizing in the cement plants to generate electricity.

Second, the potential for electricity generation from the waste heat recovery of the pyro-processing system's exhaust gases of three cement plants, which are located in Nigeria, Turkey, and Iran was investigated. The amount of electricity production depends on temperature and mass flow rate of exhaust gases, material moisture, and WHR unit efficiency. The results showed that the potentials of net electricity generation for ORC unit in plants 1, 2 and 3 are 56.1, 37.8 and 51.5 kW/t respectively, while these values for SRC unit in plant 1, 2 and 3 are 52.8, 29.8 and 43.2 kW/t, respectively.

Third, the economic feasibility of generating electricity using the WHR unit in these cement plants was discussed. The capital return rate significantly depends on the type of technology, capacity, and scale of the

WHR unit, annual kiln working hours, electricity price, and tax rate. The capital return rate for plant 3 is lower than for plants 1 and 2 because the electricity price in Iran is very low; therefore, it seems that installing the WHR unit in plants 1 and 2 is more economical than in plant 3.

## REFERENCES

[1] Gorbatenko, Y., B. Hedman, J. V. Shah, A. Sharabaroff. Waste heat recovery for the cement sector market and supplier analysis. *Institute for Industrial Productivity,* (2014).

[2] Costantino, P. D., C. D. Friedman, K. Jones, L. C. Chow, G. A. Sisson. Experimental hydroxyapatite cement cranioplasty. *Plastic and reconstructive surgery,* 90 (1992) 174-185.

[3] Taylor. H. F. *Cement chemistry*. Thomas Telford, (1997).

[4] Swamy. R. N. *Cement replacement materials*. Sheffield: Surrey University Press, (1986).

[5] Monshi, M. K. Asgarani. Producing Portland cement from iron and steel slags and limestone. *Cement and Concrete Research*, 29 (1999) 1373-1377.

[6] Kääntee, U., R. Zevenhoven, R. Backman, M. Hupa. Cement manufacturing using alternative fuels and the advantages of process modelling. *Fuel Processing Technology,* 85 (2004) 293-301.

[7] Habraken, W. J., Z. Zhang, J. G. Wolke, D. W. Grijpma, A. G. Mikos, J. Feijen, J. A. Jansen. Introduction of enzymatically degradable poly (trimethylene carbonate) microspheres into an injectable calcium phosphate cement. *Biomaterials,* 29 (2008) 2464-2476.

[8] Rasul, M. G., W. Widianto, B. Mohanty. Assessment of the thermal performance and energy conservation opportunities of a cement industry in Indonesia. *Applied thermal engineering*, 25 (2005) 2950-2965.

[9] Rodrigues, F. A., I. Joekes. Cement industry: sustainability, challenges and perspectives. *Environmental Chemistry Letters,* 9 (2011) 151-166.

[10] Kühn. K. D. What is Bone Cement?. In *The well-cemented total hip arthroplasty*, Springer, Berlin, Heidelberg, (2005) 52-59.

[11] Hwang, Y. C., D. H. Kim, I. N. Hwang, S. J. Song, Y. J. Park, J. T. Koh,, H. H. Son, W. M. Oh. Chemical constitution, physical properties, and biocompatibility of experimentally manufactured Portland cement. *Journal of endodontics*, 37 (2011) 58-62.

[12] Kordkheili, H. Y., S. Hiziroglu, M. Farsi. Some of the physical and mechanical properties of cement composites manufactured from carbon nanotubes and bagasse fiber. *Materials & Design*, 33 (2012) 395-398.

[13] Sutigno. P. Effect of aqueous extraction of wood-wool on the properties of wood-wool cement board manufactured from teak (Tectonagrandis). In *Aciar Proceedings*, (1998) 24-28.

[14] Matschei, T., B. Lothenbach, F. P. Glasser,. The AFm phase in Portland cement. *Cement and Concrete Research*, 37 (2007) 118-130.

[15] Atkins, M., F. P. Glasser, A. Kindness. Cement hydrate phase: solubility at 25 C. *Cement and concrete research*, 22 (1992) 241-246.

[16] Zhang, H., F. Xing, H. Z. Cui, D. Z. Chen, X. Ouyang, S. Z. Xu, J. X. Wang, Y. T. Huang, J. D. Zuo, J. N. Tang. A novel phase-change cement composite for thermal energy storage: Fabrication, thermal and mechanical properties. *Applied energy*, 170 (2016) 130-139.

[17] Sharma, S., N. C. Kothiyal. Influence of graphene oxide as dispersed phase in cement mortar matrix in defining the crystal patterns of cement hydrates and its effect on mechanical, microstructural and crystallization properties. *RSC Advances*, 5 (2015) 52642-52657.

[18] Sakiridis, P. E., G. D. Papadimitriou, S. Tsivilis, C. Koroneos. Utilization of steel slag for Portland cement clinker production. *Journal of Hazardous Materials*, 152 (2008) 805-811.

[19] Maki, K. Kato. Phase identification of alite in Portland cement clinker. *Cement and Concrete Research,* 12 (1982) 93-100.

[20] Li, X., W. Xu, S. Wang, M. Tang, X. Shen, Effect of $SO_3$ and MgO on Portland cement clinker: Formation of clinker phases and alite polymorphism. *Construction and Building Materials*, 58 (2014) 182-192.

[21] Kakali, G., V. Kasselouri, G. Parissakis,. Investigation of the effect of Mo, Nb, W and Zr oxides on the formation of Portland cement clinker. *Cement and Concrete Research*, 20 (1990) 131-138.

[22] Bell. L. E. Cooling, heating, generating power, and recovering waste heat with thermoelectric systems. *Science*, 321 (2008) 1457-1461.

[23] Liu, B. T., K. H. Chien, C. C. Wang,. Effect of working fluids on organic Rankine cycle for waste heat recovery. *Energy,* 29 (2004) 1207-1217.

[24] Hung, T. C., T. Y. Shai, S. K. Wang. A review of organic Rankine cycles (ORCs) for the recovery of low-grade waste heat. *Energy,* 22 (1997) 661-667.

[25] Quoilin, S., S. Declaye, B. F. Tchanche, V. Lemort. Thermo-economic optimization of waste heat recovery Organic Rankine Cycles. *Applied thermal engineering*, 31 (2011) 2885-2893.

[26] Dai, Y., J. Wang, L. Gao. Parametric optimization and comparative study of organic Rankine cycle (ORC) for low grade waste heat recovery. *Energy Conversion and Management,* 50 (2009) 576-582.

[27] Wang, E. H., H. G. Zhang, B. Y. Fan, M. G. Ouyang, Y. Zhao, Q. H. Mu. Study of working fluid selection of organic Rankine cycle (ORC) for engine waste heat recovery. *Energy*, 36 (2011) 3406-3418.

[28] Quoilin, S., R. Aumann, A. Grill, A. Schuster, V. Lemort, H. Spliethoff, Dynamic modeling and optimal control strategy of waste heat recovery Organic Rankine Cycles. *Applied energy*, 88 (2011) 2183-2190.

[29] Sprouse, C., C. Depcik. Review of organic Rankine cycles for internal combustion engine exhaust waste heat recovery. *Applied thermal engineering,* 51 (2013) 711-722.

[30] Ernst, T. C., C. R. Nelson. Cummins Inc, Waste heat recovery system with constant power output. *U.S. Patent* 7, (2011) 866, 157.

[31] Butcher, C. J., B. V. Reddy,. Second law analysis of a waste heat recovery based power generation system. *International Journal of Heat and Mass Transfer*, 50 (2007) 2355-2363.

[32] Mirolli. M. D. The Kalina cycle for cement kiln waste heat recovery power plants. *In Conference Record Cement Industry Technical Conference, IEEE,* (2005) 330-336.

[33] Campana, F., M. Bianchi, L. Branchini, A. De Pascale, A. Peretto, M. Baresi, A. Fermi, N. Rossetti, R. Vescovo. ORC waste heat recovery in European energy intensive industries: Energy and GHG savings. *Energy Conversion and Management*, 76 (2013) 244-252.

[34] Wang, J., Y. Dai, L. Gao,. Exergy analyses and parametric optimizations for different cogeneration power plants in cement industry. *Applied Energy*, 86 (2009) 941-948.

[35] Engin, T., V. Ari. Energy auditing and recovery for dry type cement rotary kiln systems: A case study. *Energy conversion and management*, 46 (2005) 551-562.

[36] Madlool, N. A., R. Saidur, M. S. Hossain, N. A. Rahim. A critical review on energy use and savings in the cement industries. *Renewable and Sustainable Energy Reviews*, 15 (2011) 2042-2060.

[37] Amiri, M. R. Vaseghi. Waste heat recovery power generation systems for cement production process. *IEEE Transactions on Industry Applications,* 51 (2014) 13-19.

[38] Oluleye, G., M. Jobson, R. Smith, S. J. Perry. Evaluating the potential of process sites for waste heat recovery. *Applied Energy*, 161 (2016) 627-646.

[39] Brueckner, S., L. Miró, L. F. Cabeza, M. Pehnt, E. Laevemann. Methods to estimate the industrial waste heat potential of regions–A categorization and literature review. *Renewable and Sustainable Energy Reviews*, 38 (2014) 164-171.

[40] Stijepovic, V. Z., P. Linke, M. Z. Stijepovic, M. L. Kijevčanin, S. Šerbanović. Targeting and design of industrial zone waste heat reuse for combined heat and power generation. *Energy,* 47 (2012) 302-313.

[41] Al-Rabghi, O. M., M. Beirutty, M. Akyurt, Y. Najjar, T. Alp,. Recovery and utilization of waste heat. *Heat Recovery Systems and CHP*, 13 (1993) 463-470.

[42] Ghalandari, V., M. M. Majd, A. Golestanian,. Energy audit for pyro-processing unit of a new generation cement plant and feasibility study for recovering waste heat: A case study. *Energy,* 173 (2019) 833-843.

[43] Ishaq, H., I. Dincer, G. F. Naterer,. Exergy and cost analyses of waste heat recovery from furnace cement slag for clean hydrogen production. *Energy,* 172 (2019) 1243-1253.

[44] Karellas, S., A. D. Leontaritis, G. Panousis, E. Bellos, E. Kakaras,. Energetic and exergetic analysis of waste heat recovery systems in the cement industry. *Energy,* 58 (2013)147-156.

[45] Song, D., L. Lin, Y. Wu. Extended exergy accounting for a typical cement industry in China. *Energy*, 174 (2019) 678-686.

[46] Colonna, P., E. Casati, C. Trapp, T. Mathijssen, J. Larjola, T. Turunen-Saaresti, A. Uusitalo,. Organic Rankine cycle power systems: from the concept to current technology, applications, and an outlook to the future. *Journal of Engineering for Gas Turbines and Power*, 137 (2015) 100801.

[47] Singh, D. V., E. Pedersen. A review of waste heat recovery technologies for maritime applications. *Energy Conversion and Management,* 111 (2016) 315-328.

[48] Kaşka,. Ö. Energy and exergy analysis of an organic Rankine for power generation from waste heat recovery in steel industry. *Energy Conversion and Management*, 77 (2014) 108-117.

[49] Zhai, H., Q. An, L. Shi, V. Lemort, S. Quoilin. Categorization and analysis of heat sources for organic Rankine cycle systems. *Renewable and Sustainable Energy Reviews*, 64 (2016) 790-805.

[50] Ishaq, H., I. Dincer. Investigation of an integrated system with industrial thermal management options for carbon emission reduction and hydrogen and ammonia production. *International Journal of Hydrogen Energy,* 44 (2019) 12971-12982.

[51] Ishaq, H. and I. Dincer. Exergy analysis and performance evaluation of a newly developed integrated energy system for quenchable generation. *Energy,* 179, (2019) 1191-1204.

[52] Kabir, G., A. I. Abubakar, U. A. El-Nafaty, Energy audit and conservation opportunities for pyroprocessing unit of a typical dry process cement plant, *Energy,* 35 (2010) 1237-1243.

[53] Fellaou, S., T. Bounahmidi, Evaluation of energy efficiency opportunities of a typical Moroccan cement plant: Part I. Energy analysis, *Applied Thermal Engineering,* 115 (2017) 1161-1172.

[54] Yang, S. C., T. C. Hung, Y. Q. Feng, C. J. Wu, K. W. Wong, K. C. Hung, Experimental investigation on a 3 kW organic Rankine cycle for low-grade waste heat under different operation parameters, *Applied Thermal Engineering*, 113 (2017) 756-764.

[55] Amiri, M. R. Vaseghi, Waste Heat Recovery Power Generation Systems for Cement Production Process, *IEEE Transactions on Industry Applications*, 51 (2015) 13-19.

[56] Ghalandari, V., H. Hashemipour, Mathematical modeling and simulation of direct reduction of iron ore in a moving bed reactor by the single particle model, *Chem. Chem. Technol.* 13 (2) (2019) 205-211.

# INDEX

## A



## B



## C